P9-CDT-239

EVERY MAN A HERO

EVERY MAN A HERO

A MEMOIR OF D-DAY, THE FIRST WAVE AT OMAHA BEACH, AND A WORLD AT WAR

RAY LAMBERT

AND JIM DEFELICE

wm

WILLIAM MORROW

An Imprint of HarperCollins*Publishers*

 For information, address HarperCollins Publishers, 195 Broadway, New York, NY 10007.

HarperCollins books may be purchased for educational, business, or sales promotional use. For information, please email the Special Markets Department at SPsales@harper collins.com.

FIRST EDITION

Maps courtesy of the Department of History, U.S. Military Academy at West Point.

Library of Congress Cataloging-in-Publication Data has been applied for.

ISBN 978-0-06-293748-3

19 20 21 22 23 DIX/LSC 10 9 8 7 6 5 4 3

Dedicated to those I've lost—Bill, Estelle,
my medics, my good friends . . .

And to those who help me remember—Arthur, Linda,
Barbara, my family & my good friends.

Almighty God: Our sons, pride of our Nation, this day have set upon a mighty endeavor, a struggle to preserve our Republic, our religion, and our civilization, and to set free a suffering humanity.

Lead them straight and true; give strength to their arms, stoutness to their hearts, steadfastness in their faith.

They will need Thy blessings. Their road will be long and hard. For the enemy is strong. He may hurl back our forces. Success may not come with rushing speed, but we shall return again and again; and we know that by Thy grace, and by the righteousness of our cause, our sons will triumph.

They will be sore tried, by night and by day, without rest—until the victory is won. The darkness will be rent by noise and flame. Men's souls will be shaken with the violences of war.

For these men are lately drawn from the ways of peace. They fight not for the lust of conquest. They fight to end conquest. They fight to liberate. They fight to let justice arise, and tolerance and good will among all Thy people. They yearn but for the end of battle, for their return to the haven of home.

Some will never return. Embrace these, Father, and receive them, Thy heroic servants, into Thy kingdom.

And for us at home—fathers, mothers, children, wives, sisters, and brothers of brave men overseas, whose thoughts and prayers are ever with them—help us, Almighty God, to rededicate ourselves in renewed faith in Thee in this hour of great sacrifice.

—FRANKLIN D. ROOSEVELT, D-DAY PRAYER, JUNE 6, 1944

Contents

List of Maps

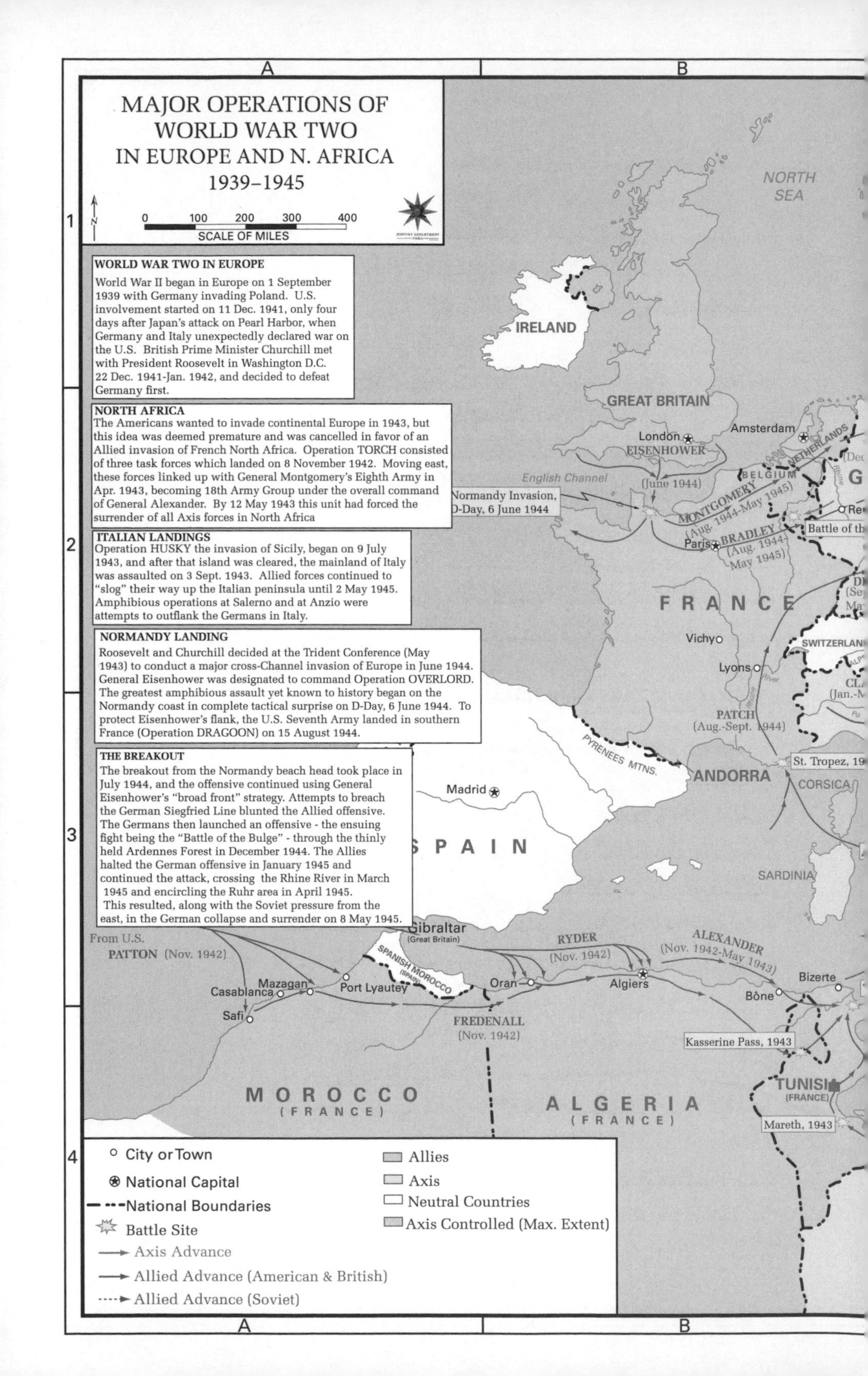
MAJOR OPERATIONS OF WORLD WAR TWO IN EUROPE AND N. AFRICA 1939–1945
0 100 200 300 400
SCALE OF MILES
WORLD WAR TWO IN EUROPE
World War II began in Europe on 1 September 1939 with Germany invading Poland. U.S. involvement started on 11 Dec. 1941, only four days after Japan's attack on Pearl Harbor, when Germany and Italy unexpectedly declared war on the U.S. British Prime Minister Churchill met with President Roosevelt in Washington D.C. 22 Dec. 1941-Jan. 1942, and decided to defeat Germany first.
NORTH AFRICA
The Americans wanted to invade continental Europe in 1943, but this idea was deemed premature and was cancelled in favor of an Allied invasion of French North Africa. Operation TORCH consisted of three task forces which landed on 8 November 1942. Moving east, these forces linked up with General Montgomery's Eighth Army in Apr. 1943, becoming 18th Army Group under the overall command of General Alexander. By 12 May 1943 this unit had forced the surrender of all Axis forces in North Africa
ITALIAN LANDINGS
Operation HUSKY the invasion of Sicily, began on 9 July 1943, and after that island was cleared, the mainland of Italy was assaulted on 3 Sept. 1943. Allied forces continued to "slog" their way up the Italian peninsula until 2 May 1945. Amphibious operations at Salerno and at Anzio were attempts to outflank the Germans in Italy.
NORMANDY LANDING
Roosevelt and Churchill decided at the Trident Conference (May 1943) to conduct a major cross-Channel invasion of Europe in June 1944. General Eisenhower was designated to command Operation OVERLORD. The greatest amphibious assault yet known to history began on the Normandy coast in complete tactical surprise on D-Day, 6 June 1944. To protect Eisenhower's flank, the U.S. Seventh Army landed in southern France (Operation DRAGOON) on 15 August 1944.
THE BREAKOUT
The breakout from the Normandy beach head took place in July 1944, and the offensive continued using General Eisenhower's "broad front" strategy. Attempts to breach the German Siegfried Line blunted the Allied offensive. The Germans then launched an offensive - the ensuing fight being the "Battle of the Bulge" - through the thinly held Ardennes Forest in December 1944. The Allies halted the German offensive in January 1945 and continued the attack, crossing the Rhine River in March 1945 and encircling the Ruhr area in April 1945. This resulted, along with the Soviet pressure from the east, in the German collapse and surrender on 8 May 1945.
NORTH SEA
IRELAND
GREAT BRITAIN
London
EISENHOWER
(June 1944)
Amsterdam
NETHERLANDS
BELGIUM
English Channel
Normandy Invasion, D-Day, 6 June 1944
MONTGOMERY (Aug. 1944-May 1945)
BRADLEY (Aug. 1944-May 1945)
Paris
Battle of th
FRANCE
Vichy
Lyons
SWITZERLAN
PATCH (Aug.-Sept. 1944)
PYRENEES MTNS.
ANDORRA
St. Tropez, 19
CORSICA
Madrid
SPAIN
SARDINIA
Gibraltar (Great Britain)
From U.S.
PATTON (Nov. 1942)
SPANISH MOROCCO (SPAIN)
Casablanca
Mazagan
Port Lyautey
Safi
RYDER (Nov. 1942)
Oran
Algiers
ALEXANDER (Nov. 1942-May 1943)
Bizerte
Bône
FREDENALL (Nov. 1942)
Kasserine Pass, 1943
TUNISIA (FRANCE)
Mareth, 1943
MOROCCO (FRANCE)
ALGERIA (FRANCE)
City or Town
National Capital
National Boundaries
Battle Site
Axis Advance
Allied Advance (American & British)
Allied Advance (Soviet)
Allies
Axis
Neutral Countries
Axis Controlled (Max. Extent)

A
C
D
1
2
3
4
FINLAND
SWEDEN
Oslo
Helsinki
Leningrad
Stockholm
(Jan.-Dec. 1944)
ESTONIA
BALTIC SEA
Riga
LATVIA
LITHUANIA
Memel
Copenhagen
Smolensk
(June 1944-Feb. 1945)
EAST PRUSSIA (GER.)
Danzig (Gdansk)
Minsk
SOVIET UNION
Gomel
Kursk, 1943
Berlin
Warsaw
Brest
(July 1943-Dec. 1944)
(Dec. 1944-May 1945)
Torgau
POLAND
(July 1943-June 1944)
Kharkov
Kiev
(June 1944-May 1945)
Prague
Auschweitz
Lvov
CZECHOSLOVAKIA
CARPATHIAN MOUNTAINS
Vienna
AUSTRIA
Budapest
HUNGARY
Trieste
(June 1944-May 1945)
ROMANIA
Sevastopol
Yalta
Belgrade
Bucharest
Danube River
YUGOSLAVIA
BLACK SEA
ADRIATIC SEA
BULGARIA
ITALY
Sofia
ALBANIA
Istanbul
Naples
Salerno, 1943
TURKEY
GREECE
AEGEAN SEA
MALTA (GREAT BRITAIN)
DODECANESE ISLAND (ITALY)
CYPRUS (GREAT BRITAIN)
CRETE
MEDITERRANEAN SEA
Benghazi
Tobruk, 1942
Cairo
LIBYA (ITALY)
EGYPT
River
RED SEA

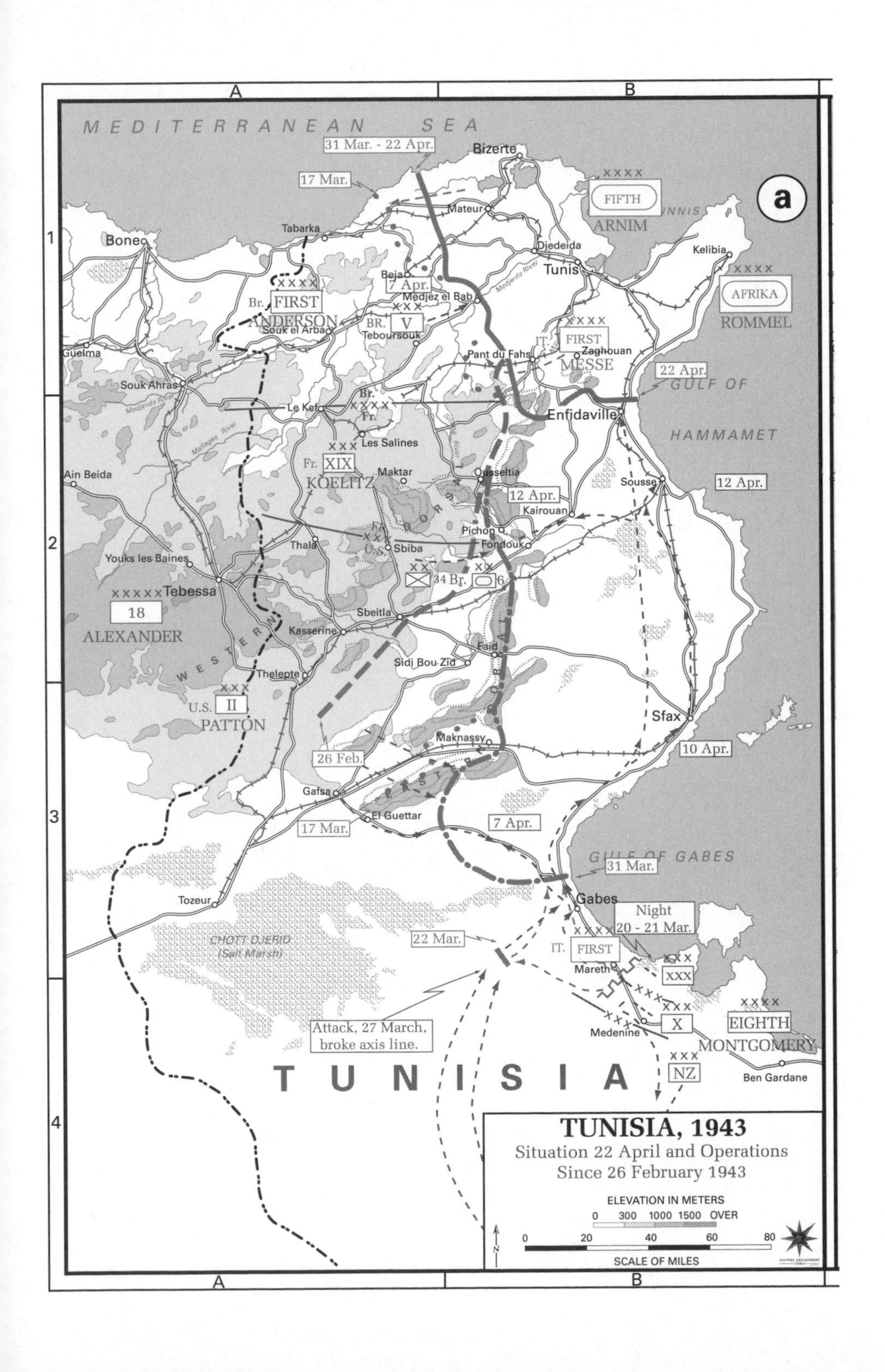

MEDITERRANEAN SEA
31 Mar. - 22 Apr.
Bizerte
17 Mar.
FIFTH
ARNIM
a
Mateur
Tabarka
Bone
Djedeida
Tunis
Kelibia
Beja
7 Apr.
Medjez el Bab
Br. FIRST
ANDERSON
BR. V
Souk el Arba
Teboursouk
AFRIKA
ROMMEL
IT. FIRST
MESSE
Zaghouan
Pant du Fahs
Guelma
Souk Ahras
22 Apr.
GULF OF
HAMMAMET
Le Kef
Enfidaville
Les Salines
Fr. XIX
KOELITZ
Maktar
Ousseltia
Sousse
12 Apr.
Ain Beida
12 Apr.
Kairouan
Pichon
Fondouk
Thala
Sbiba
Youks les Baines
34 Br.
6
Tebessa
18
ALEXANDER
Sbeitla
Kasserine
Faid
Sidi Bou Zid
Thelepte
U.S. II
PATTON
Sfax
Maknassy
26 Feb.
10 Apr.
Gafsa
El Guettar
17 Mar.
7 Apr.
GULF OF GABES
31 Mar.
Gabes
Tozeur
Night 20 - 21 Mar.
CHOTT DJERID (Salt Marsh)
22 Mar.
IT. FIRST
Mareth
XXX
X
EIGHTH
MONTGOMERY
Medenine
Attack, 27 March, broke axis line.
TUNISIA
NZ
Ben Gardane
TUNISIA, 1943
Situation 22 April and Operations Since 26 February 1943
ELEVATION IN METERS
0 300 1000 1500 OVER
0 20 40 60 80
SCALE OF MILES

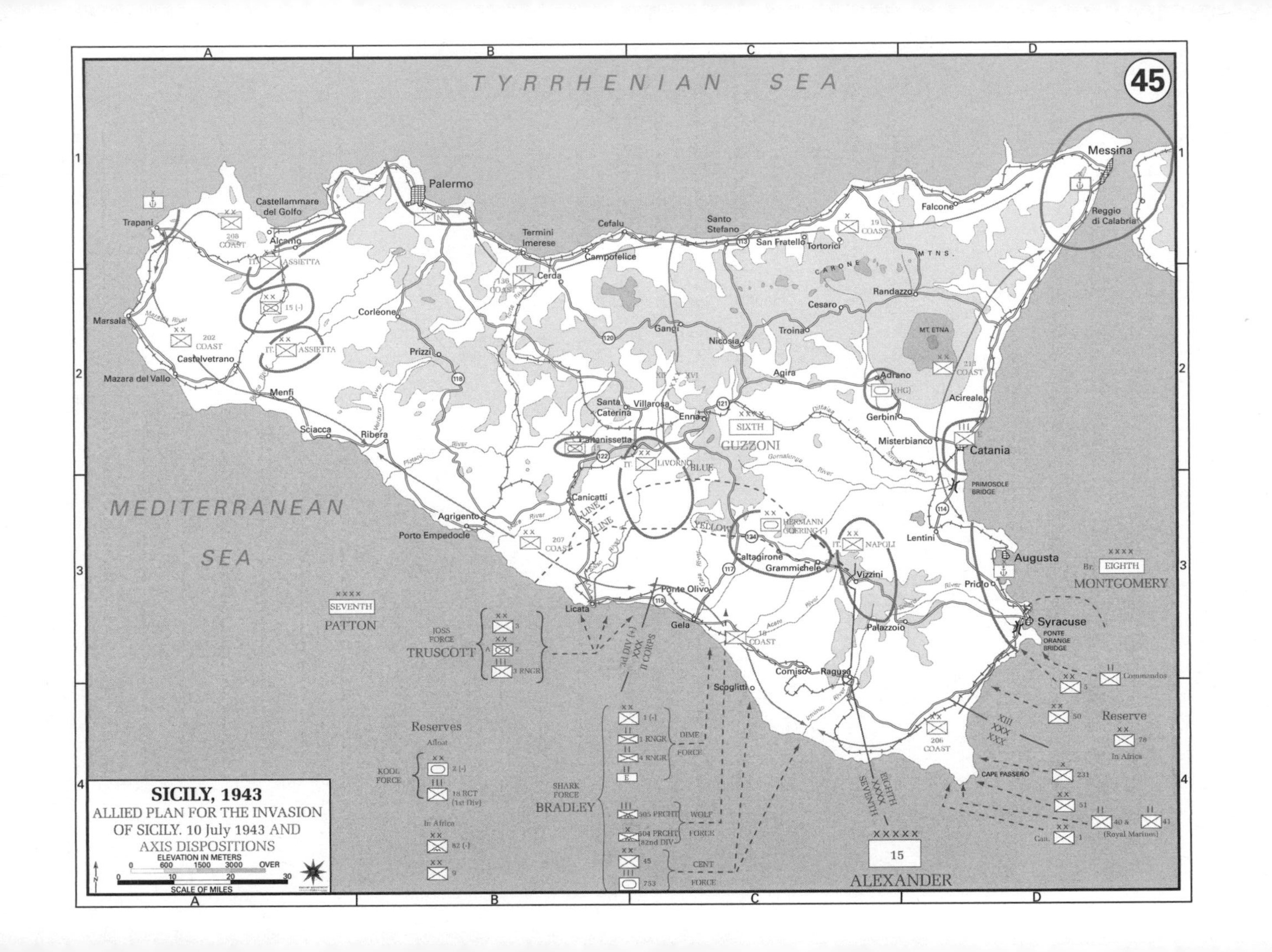

SICILY, 1943
ALLIED PLAN FOR THE INVASION OF SICILY, 10 July 1943 AND AXIS DISPOSITIONS

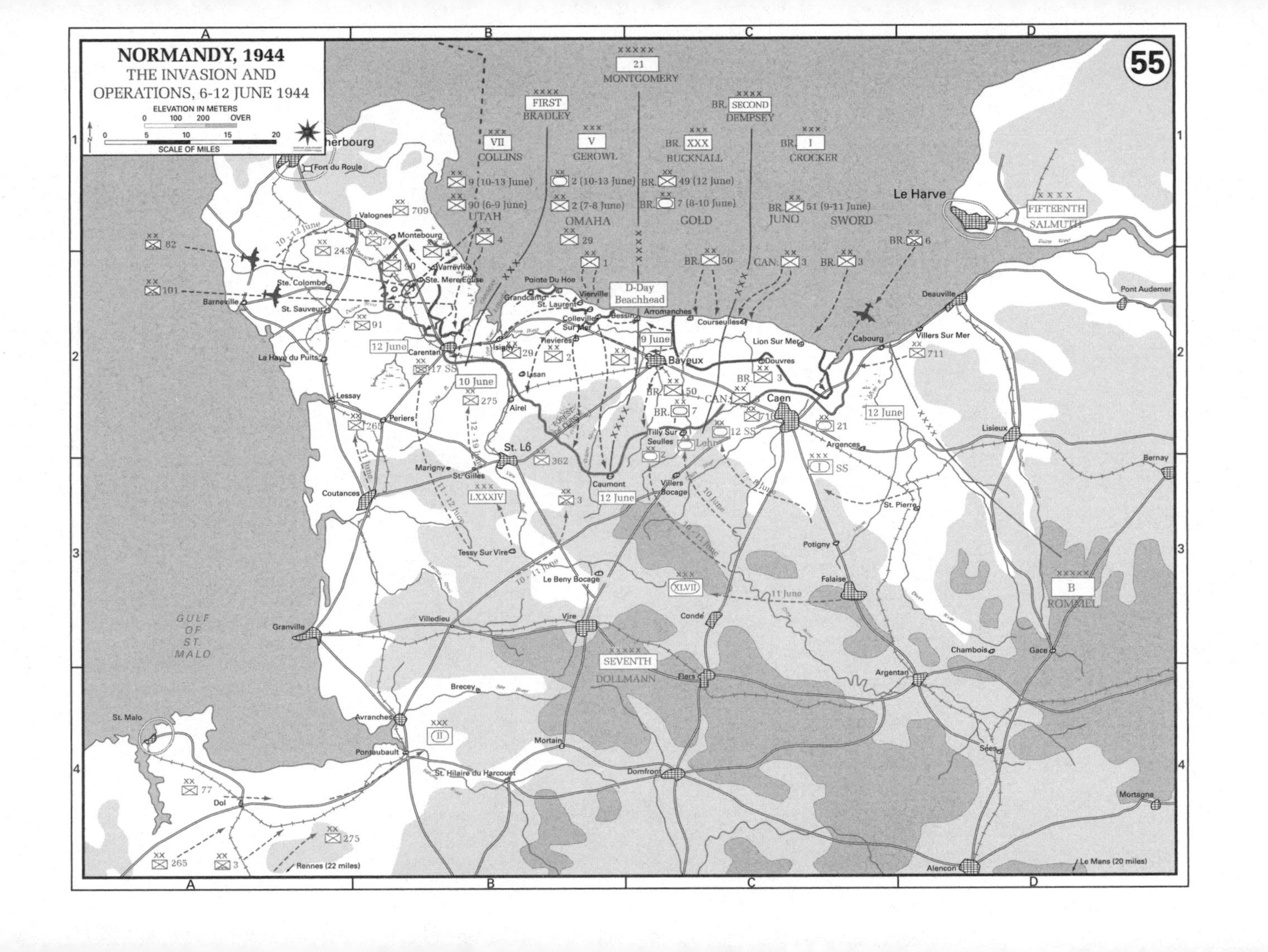
55
NORMANDY, 1944
THE INVASION AND
OPERATIONS, 6-12 JUNE 1944
ELEVATION IN METERS
0 100 200 OVER
0 5 10 15 20
SCALE OF MILES
21
MONTGOMERY
FIRST
BRADLEY
BR. SECOND
DEMPSEY
VII
COLLINS
V
GEROWL
BR. XXX
BUCKNALL
BR. I
CROCKER
9 (10-13 June)
90 (6-9 June)
UTAH
2 (10-13 June)
2 (7-8 June)
OMAHA
BR. 49 (12 June)
BR. 7 (8-10 June)
GOLD
BR. 51 (9-11 June)
JUNO
SWORD
FIFTEENTH
SALMUTH
Le Harve
D-Day
Beachhead
BR. 50
CAN. 3
BR. 3
BR. 6
82
101
709
243
91
17 SS
275
265
362
711
21
12 SS
Lehr
LXXXIV
I SS
XLVII
SEVENTH
DOLLMANN
B
ROMMEL
II
77
265
3
275
Cherbourg
Fort du Roule
Valognes
Montebourg
Varreville
Ste. Mere Eglise
Ste. Colombe
Barneville
St. Sauveur
La Haye du Puits
Carentan
Isigny
Grandcamp
Pointe Du Hoe
Vierville
St. Laurent
Colleville Sur Mer
Rievieres
Bessin
Arromanches
Courseulles
Lion Sur Mer
Douvres
Bayeux
Caen
Lisan
Airel
Lessay
Periers
Marigny
St. Gilles
St. Lô
Caumont
Tilly Sur Seulles
Villers Bocage
Coutances
Tessy Sur Vire
Le Beny Bocage
Villedieu
Vire
Granville
Brecey
Avranches
Pontaubault
St. Hilaire du Harcouet
Mortain
Domfront
Flers
Conde
Falaise
Potigny
Argences
St. Pierre
Cabourg
Villers Sur Mer
Deauville
Pont Audemer
Lisieux
Bernay
Chambois
Gace
Argentan
Sées
Mortagne
Alencon
Le Mans (20 miles)
Rennes (22 miles)
St. Malo
Dol
GULF OF ST. MALO
12 June
10 June
9 June
10-12 June
12-19 June
11-12 June
11 June
7-8 June
10 June
10-11 June
A
B
C
D
1
2
3
4

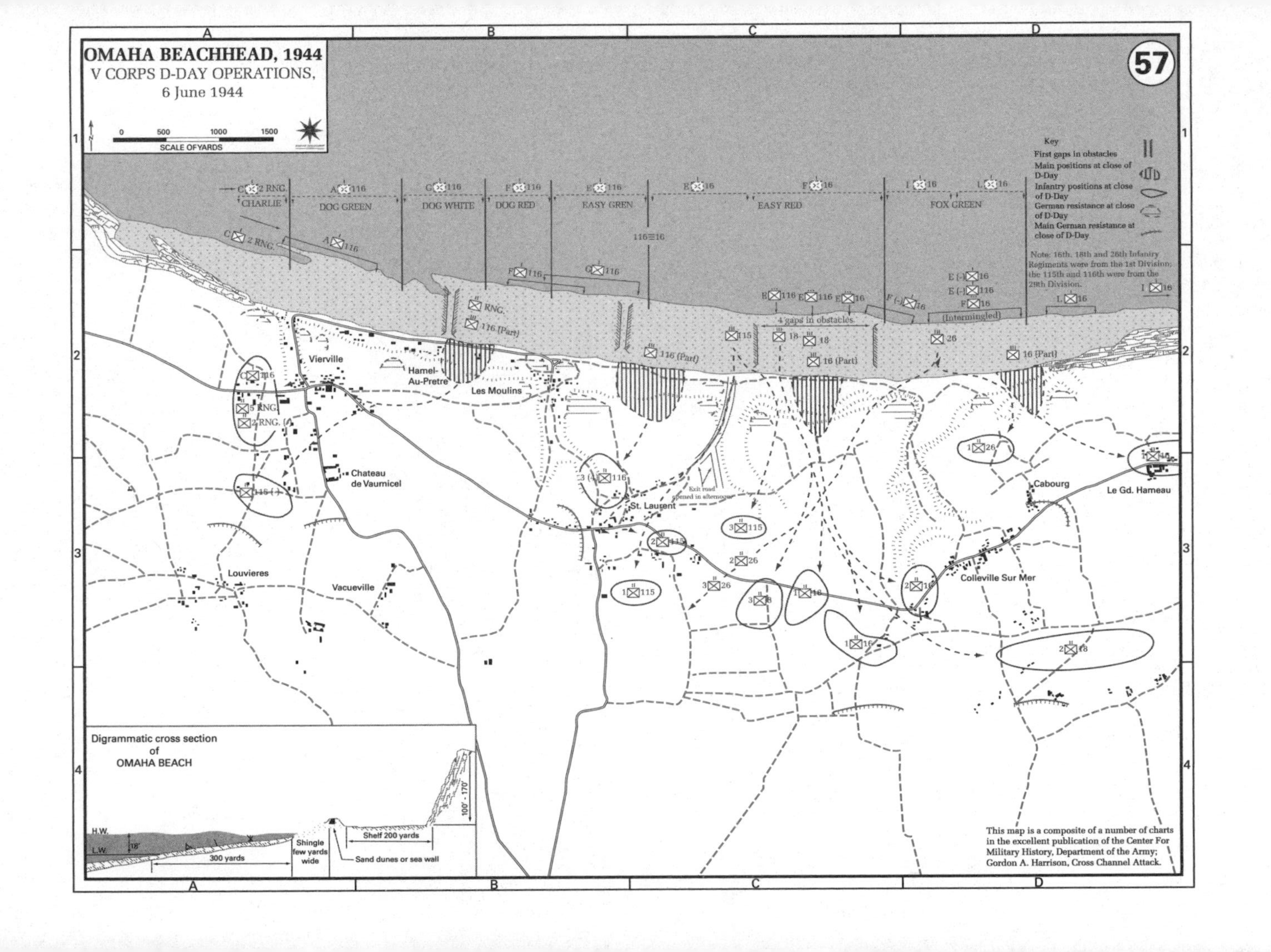
OMAHA BEACHHEAD, 1944
V CORPS D-DAY OPERATIONS,
6 June 1944
57
SCALE OF YARDS
0 500 1000 1500
Key
First gaps in obstacles
Main positions at close of D-Day
Infantry positions at close of D-Day
German resistance at close of D-Day
Main German resistance at close of D-Day
Note: 16th, 18th and 26th Infantry Regiments were from the 1st Division; the 115th and 116th were from the 29th Division.
CHARLIE
DOG GREEN
DOG WHITE
DOG RED
EASY GREEN
EASY RED
FOX GREEN
4 gaps in obstacles
(Intermingled)
116 (Part)
16 (Part)
Exit road opened in afternoon
Vierville
Hamel-Au-Pretre
Les Moulins
Chateau de Vaumicel
St. Laurent
Cabourg
Le Gd. Hameau
Colleville Sur Mer
Louvieres
Vacueville
Digrammatic cross section of OMAHA BEACH
H.W.
L.W.
300 yards
Shingle few yards wide
Sand dunes or sea wall
Shelf 200 yards
100' - 170'
This map is a composite of a number of charts in the excellent publication of the Center For Military History, Department of the Army; Gordon A. Harrison, Cross Channel Attack.

Introduction:
O-Hour+15

THAT DAY, 75 YEARS AGO

By Jim DeFelice

0645, 6 June 1944. D-Day.

OMAHA BEACH, NORMANDY, FRANCE

Imagine you are one among 160,000, about to join the greatest battle of the twentieth century.

The sun has been up for well over an hour, but you haven't seen it, partly because it's blocked by a shroud of thick clouds straining to hold back rain. The bigger reason is this: you have neither the energy nor the space to raise your head, let alone the will, for to look up is to break the spell keeping you safe.

The spell is an illusion, and you know it. Yet you cling to it as firmly as you can, gripping it harder even than your well-wrapped M1 rifle, as you shift uncomfortably in the landing craft. You're hurtling toward Omaha Beach in Normandy, France. Dark smoke already covers your target. The whole world seems black, except the red flames from the burning boats nearby.

You've dreaded and prayed for this day to arrive for weeks. At turns you've been stoic, ambivalent, confident, fearful. Now it's finally here, and those emotions and twenty more are exploding inside you, threatening to pierce the thin shell of your psyche. Around you bullets ricochet off the hull of the boat.

For all the landings you've practiced, there is no model for this, no precedent, no preconception to force the shattered kaleidoscope of chaotic reality surrounding you into an ordered outline.

You cannot hear distinct sounds. The engines, the shells, the gunfire—they've blurred into a roaring mix, half-thunder, half-symphony, orchestrated by distant, angry gods.

The landing craft stops. The ramp splashes down. People shout, "Go!"

You blink your eyes and try to stir, only to realize you're already moving, propelled forward by a mysterious momentum, not by courage or duty or even will. Two steps onto the ramp and you're now half-swimming. You've been let off in deeper water than you thought, farther away from the beach, but not the danger. The only direction is forward—toward something both more violent and more epic than you've ever experienced.

Your brain is clogged with a thousand competing thoughts, most of them useless, some paralyzing.

But one rises above all others:

Who will save me if I am hit?

The answer is already ashore, plunging and wading and pushing against the wind and wild waves amid mortar shells and gunfire.

He has saved dozens of others this momentous day: Ray Lambert, an army medic from Alabama. By the time the ramp on your Higgins boat goes down, he'll have been doing this for nearly a half hour.

Staff Sergeant Arnold Raymond "Ray" Lambert is twenty-three, an old man by the standards of this battle. He's in charge of a medical team, and this is his third seaborne invasion. He previously saw action in Africa and Sicily. As difficult as those battles were—and they were among the worst of the Second World War—Normandy is a different hell. By the end of the day, some two thousand men will be wounded or killed on Omaha Beach; many of those will die in the wildly misnamed "Easy Red," a small rectangle of sand targeted by the 16th Infantry. The fighting on Omaha will be the most horrific of the invasion, so bad that the general in charge will think seriously of retreating—a development that could bring disaster to the other beaches and the entire operation, perhaps even the war.

It's Ray's job to stave off defeat by helping as many men as possible. In some ways, he was a born soldier, hunting from an early age in a time and place where it was a means of survival rather than a hobby. As a teenager, he carried a pistol in his belt to deal with unruly farmers trying to keep him from doing his job as a county veterinarian. He spent summer months chopping down trees for a lumber business. He has a well-tested middleweight's hook, honed in army boxing rings as well as Alabama farm country.

But this fighter was trained by the army to be a medic. Selected almost by accident, he has learned the art as well as the science of battlefield medicine.

The most important thing: make sure the infantrymen know you're there.

That means Ray can't hide when the bullets fly. He can't dig a foxhole. He can't retreat. He can't think of himself, but rather the men he must save.

By this point in the war, he's won Silver and Bronze Stars and two Purple Hearts. He may be the only medic in the army with a sharpshooter's badge.

At the moment, Ray Lambert is not thinking of any of that. He hit the beach twenty minutes ago aiming to establish a station where the wounded could be triaged and receive first aid. But it was evident even before he left the landing craft that no aid station could be set up on Easy Red for quite a while.

Now Ray is looking, scanning the water for men who have been hit and can't make it to shore on their own.

One is clinging to an obstruction.

Foolish—the German weapons are zeroed in on each. Stay there and die.

Ray leaps toward him. Weighed down by sodden clothes and his med bags, the medic struggles hard against the waves. He reaches the man, yells but can't be heard.

"We're going to the beach!"

He tugs, but the man doesn't move. Belatedly, Ray realizes the man is caught on barbed wire below.

He dives down, tries to pull him loose, then unhook him. The salt water stings his eyes. He resurfaces for air, keeping his head low to avoid the bullets flying overhead.

Ray pulls again. When that doesn't work, he dives below once more. He sees a snagged strap and unhooks it, resurfaces, then by some miracle yanks the GI free.

The soldier's head slides back and sinks beneath the waves. His rucksack is so heavy that it counteracts the life belt he's wearing.

Ray gets him upright. Together they start toward the beach, and the only safe place Ray has found: a slab of rock, perhaps the remains of a concrete bunker, that stands like a thumb on the beach ahead of the obstructions.

They move toward shore. The noise is so loud that eardrums shatter. People go down in front of them, but Ray knows that they can't stop or they will be swept back by the waves, or worse, under them, back to the worst of the mines and booby traps, wire and steel that lay ready to slice or blow them open.

At last, they make it to the row of bodies floating in the bubbled surface at water's edge. Ray pushes the man along, hoping he doesn't see.

Finally, they reach the rock. Ray takes a minute, then begins checking the man's wounds.

This one, he thinks, will live. If he stays behind the rock.

Ray Lambert will be seriously injured minutes later. But those wounds will not stop him. Nothing will, until he's saved several more lives. Then his back will be broken so severely he will lose consciousness. And in an ironic turn utterly characteristic of war, it will be an American landing craft, not a German bomb, that does him in.

Energy spent, he'll curl up behind the shattered wall of concrete where he had taken so many men before. There he will wait for whatever comes next, be it salvation or death, or both.

EVERY MAN A HERO

Prologue:
Why We Remember

THE BEACH AND THE ROCK

Colleville-sur-Mer is a picturesque village in northern France, blessed with a lovely beach on the English Channel. Take the winding road down to the water just after sunrise on a nice summer day, and chances are good you will find riders exercising their horses along the surf. Go a few hours later, and the place will be filled with families camped out in the sun, enjoying the sand and water.

I've seen it that way myself. But for me, a far different scene is never far from my mind.

In the early dawn of June 6, 1944, it was a place of death and sacrifice. For the beach below Colleville was the center of a place known to me and my companions as Omaha Beach, the bloodiest of all the beaches where the Allies landed on D-Day. Where tourists and vacationers see pleasant waves, I see the faces of drowning men. Amid the sounds of children playing, I hear the cries of men pierced by Nazi bullets. Where tall grass on the bluff wavers in the wind, I catch glimpses of GIs treading through barbed wire to turn the tide of battle.

Today, the sand and rocks are pristine. In my memory, they are stained red with blood.

This is a place where I saved more than a dozen men. It is also the place I nearly died. Above all, it is the place where courage proved that even the worst evils can be overcome. The joy on the beach today is proof.

Reminders of the battle are close at hand: A beautiful museum dedicated to the combat here. An austere yet inspiring cemetery on the bluff above. Monuments, and the remains of bunkers and embattlements, long since pacified.

And there is a rock, a mass of aggregate some six or eight feet wide and four feet high that interrupts the smooth expanse of sand below the village. In any other place, it might seem an aberration. But it is here because on that momentous day it provided shelter for the wounded.

I know that rock well. I dragged several men to it, as did my medics from the 2nd Battalion of the 16th Infantry Regiment, part of the American army's 1st Division. It was all the shelter we could find in the first hours of the assault. It has remained, a testament to the French who still celebrate their release from occupation, and a memorial to the men who gave their lives so the French might be freed.

For many years, I kept my story of that day to myself. Largely, this was because I chose to move on. The war, as life-changing as it was for me and for my entire generation, was only a part of who I was, who we were. I had a family to feed; we had a nation and a world to rebuild. But I also felt that my story was not worth sharing. I landed at Omaha, but thousands did. I fought from Africa to Sicily and then to France, but hundreds of others did the same. I was and am no better than they. Others call me a hero, but I would never use the word to describe myself.

I did what I was called to do. As a combat medic, my job was to save people, and to lead others who did the same. I was proud of that job, and remain so. But I was always an ordinary man, not one who liked being at the head of a parade.

Most of all, I am a man who looks forward, not back. Even at age ninety-eight, I have a slate of things to accomplish. Simple things, mostly—repairing a water heater to donate to a church, cleaning the yard of winter debris.

Lately, though, I have come to see that I have yet another important task, one I never wished for or imagined when I was first on that beach. My job now is to remember, not for my sake, but for the sake of others. For at ninety-eight, I am one of the last men left who was there on that day. I am one of the few remaining links to the courage and strength that carried the Allies to victory, and to the men who made June 6, 1944, D-Day, a day to remember for all time.

President Franklin D. Roosevelt called it a "Mighty Endeavor." It was that, and more. One hundred and sixty thousand men, some five thousand ships, and thirteen thousand airplanes took part in an assault that ultimately decided the war. It was one of the bloodiest days in one of the bloodiest conflicts mankind has ever fought. It ended in a tremendous victory, but one that was far from preordained.

Every man on that beach was a hero. Each one braved incredible gunfire, artillery, mortar shells, obstructions, mines. Each man had his own story.

This book tells mine. I share it not for myself, but to tell you what we all went through—and to show that whatever difficulties you, too, encounter, they can be overcome.

While we were working on this book, I found an old newspaper with a story about my medic unit from before the war. The page was brittle, the ink faded, but if you were patient and careful, the words came clear enough. It seems to me that is a metaphor not only for memory, but my aim: to pass down what I remember before it fades, so you, too, can know and remember.

Age tugs at me, dimming what I can see when I look back. But I found that working on this book sharpened what I knew, making my memory clearer. That, too, is a metaphor. The harder we work to remember, the better we get at it. The more we remember, the better we become at mastering the present.

But let me start at the beginning, so you will see that I am just an ordinary man like you.

ONE

Early Days

DOWN ON THE FARM

I like to tell people I'm older than Noah and the Flood, but that's not true. I'm not even older than America, or the farm I was born on, where the barn, still standing last time I looked, was put up some 250 years ago.

The day was November 26, 1920. There was a midwife, no doctor—doctors at births were a rare luxury in the rural farmland of Alabama where I made my debut as Arthur Raymond Lambert, the second son of William and Bessie Jane Lambert.

The farmhouse where I was born belonged to my grandparents and was located a few miles outside of Clanton, Alabama, which even today is a fairly small town. For the first years of my life, my parents rented a house nearby in Maplesville, a town even more rural and smaller than Clanton. My father and his two brothers, Alvin and Walter, worked for their dad in a lumber business. Grandfather Lambert, who had a store and other interests in

Selma, would buy a piece of land for its timber. His sons would move a sawmill there—it had a big gas-powered automobile engine to run the blade and conveyor belt—and set up the operation.

The logs didn't just waltz themselves over to the machine. Trees would be cut down, trimmed, and then transported to the sawmill. Depending on what was called for, they would be milled into shape and then transported to Grandfather's business for sale.

It was hard and sometimes dangerous work. It also meant that we moved several times while I was young, renting houses near the timber and staying there while the wood was harvested. Eventually, my father rented a piece of property big enough to farm. We kept livestock—he and my Uncle Alvin eventually owned some fifty head—and a lot of the day-to-day work caring for the animals and tending to crops like corn fell naturally to myself and my brothers.

America in 1920 was just starting to wake up to its potential as a world leader. We'd turned the tide in World War I, joining Great Britain, France, and the other Allies after three years of being tested by Germany.

We were reluctant to join that war, and maybe reluctant to recognize our responsibilities as a world power. But once we were in it, there was no turning back. America had gone through some extreme changes since World War I, becoming an industrial powerhouse and transitioning from a mostly rural nation to one where cities not only dominated the economy but held most of the population. It was the start of the Roaring Twenties, a time when, at least by legend, anything went. Mass production of cars and airplanes, the adoption of radio as a mass medium, medical advances—so many different inventions altered life on many levels. Women finally got the right to vote in America the same year I was born.

But for a lot of us, especially in the heart of Alabama where I was born and raised, things were pretty much the way they'd been since the foundation of our old barn was first put down. We had no running water, no indoor toilets, and no electricity. Air-conditioning meant opening the windows to catch a breeze.

Refrigerators were just starting to become popular home items around the time I was born, but they didn't reach most homes in rural Alabama, at least not ours, for many more years. We kept our meat by smoking or curing it—funny that meat like that would be considered a delicacy today, though you can't argue with how good it tastes. Lard—pig fat—was used in just about everything. Not the best thing for your health, according to the doctors.

Folks didn't have closets full of clothes. They might own as many as three or four pairs of shirts and overalls, one of which would be fairly new and always clean—that was what you wore to church on Sunday.

Farms were all family affairs back then, which means that the kids were part of the labor force. We were set up with chores—easy ones, of course—practically as soon as we could walk. Fetching wood or water, tending to the animals—it was routine for us and pretty much any kid who lived in rural America in those days. It taught you a lot, and not just about how to deal with an angry rooster or a temperamental tractor engine. Responsibility and an appreciation for hard work were not just ideals to strive for when you lived on a farm; they were what you had to do to survive.

I mentioned refrigerators, and I guess I shouldn't say we didn't have one. We did—but it was actually a well. You could keep your milk and butter down there, above the water, of course, retrieving it by rope. Up north you'd have maple syrup on your pancakes for breakfast; where I lived we'd raise sugarcane, press it (or have it pressed at a nearby mill), and cook it so that it became molasses. Different source, but roughly the same idea when applied to pancakes.

The answers to the questions I'm sure you're dying to ask: *If you didn't have plumbing, where did you get your water?*

We'd draw it from the well—lower a bucket with a rope and haul it back up.

Yes, we used an outhouse.

No, we did not have soft toilet paper.

Yes, any paper, including store catalogs, would do.

I can't say I have fond memories of digging the pits—a typical job for a kid—or handling the lime that you'd throw down the hole as an air freshener. But after so many years gone now, what are left are memories of happy times. I'm sure there were bruises and nicks and stumbles along the way, but they're all long healed. I remember blueberries in early summer—sweet candy, fresh off the bush. I remember my grandma packing up the kids in the wagon behind the horse. I remember what seems a perfect time in a perfect place very long ago.

I had two brothers, Euel and Harland. Euel was about two years older than me, and we spent a lot of our lives together, physically and mentally. We fought quite a bit, as brothers sometimes do, but as we got older the fights stopped and we became pretty close friends. We did chores together, went to school together, shared things only brothers share. We would sense what the other was thinking just by looking at him—a good thing, since neither one of us was what you might call a talker. Since he was my older brother, I always felt safe when he was around; he would back me up when I needed it.

I'd do the same for him, no questions asked.

Harland was two years younger than me. Harland was what I would call a free spirit, or at least more than I was. Maybe it was just that his energy put him in so much motion that he couldn't be satisfied staying still in one place too long. Cancer cut him down when he was in his mid-fifties. You don't get used to loss, not really, but you can grow philosophical about it. You realize it's going to happen to us all.

My sister, Gloria, was born many years later; during the war, in fact. That age difference was huge, especially in those days when your older siblings were all boys. I'm sure it seemed to her that her older brothers were always bossing her around, or acting more like parents than brothers. But we loved her dearly, and maybe on occasion spoiled her just a little bit. I don't think I've known a kinder, more loving person in my life.

I got my first real taste of the family lumber business the year I turned thirteen. My cousin Durwood Williamson and I got the job of cutting down the trees. Chainsaws were well off in the future—our biceps supplied the power for the two-man saws. It was man's work, and we were proud to do it, even if it had us going as long as there was sunlight.

We were old enough to cut down trees, old enough to drive the trucks that hauled the logs to the saw, but we weren't old enough to drink coffee—at least not me. Maybe the folks thought it would stunt my growth, a not uncommon belief back in the day.

And yet . . . for some reason, red-eye gravy, which is made from pan scrapings and coffee, was fine.

I grew up in the South with the two B's—if you got out of line, you got the Bible and the belt, though not usually in that order. Maybe we weren't better behaved than children now, but I would say we learned to respect our elders and mind our manners darn quick. If the world wasn't a better place for it, at a minimum it was more polite.

THE DEPRESSION

The Roaring Twenties, of course, gave way to the Great Depression, the worst financial bust our country has ever known. Living through it, there were no sharp dividing lines, no one day where prosperity ended and poverty began. Sure, there was a stock market crash—that was in October 1929—but there had been crashes before, and Wall Street was far away. It was just one factor among many that either caused or were part of the Great Depression.

We were poor, but so was everyone else. When you're a kid, making your own toys and finding some way to amuse yourself for free don't seem like a big deal if all your friends are doing it. Being hungry is something else, but in our case we were lucky to always have food.

We moved from Maplesville to Prattville when I was eight; a few years later, we relocated to Collirene, about twenty-five miles as the crow flies southwest. If the other places I'd lived were small, Collirene was tiny—a grand total of fifteen kids populated the entire elementary school. Our teacher was an angel, hardworking, caring, and so dedicated. I still remember how important she made me feel bending over my desk to check my work or answer some question I had.

We grew up fast in those days; kids handled jobs that today would be reserved for people in their late teens and beyond. Aside from learning how to handle livestock—besides the cows and chickens, we also had hogs—I'd fix machines when they broke down, and I got to be a fair mechanic. Even more impressive, at least looking back, was the summer job I got at eleven working a bulldozer. I helped build a road, and my specialty was uprooting tree stumps so the graders could smooth the bed.

The bulldozer was a mechanical beast, but not nearly as ornery as its flesh-and-blood counterparts can be. With my dad and uncle raising cattle, I was put to work helping to care for the animals, learning to rope and brand just as if I were a cowboy out west. And just like a cowboy, I learned to break horses. Another time and place, I could have been Buffalo Bill or maybe just an anonymous cowpoke, earning his living on the Plains.

High school was in Hayneville, some eighteen miles away. Unlike when I attended elementary school, I didn't have to walk; I caught the bus. Which was welcome, given that I'd been up for hours milking the cows and then setting them out to pasture. And I knew the driver—it was my brother Euel.

By this point, the Depression was in full steam, and the lack of money was evident to everyone, even young teenagers like myself. People bartered for food or clothes or other necessities. And neighbors tried to help each other. I remember butchering a cow—another skill I picked up young—and delivering the meat to people in town. We weren't selling it; we were giving it away. It was an act of charity and neighborliness. You looked out for your family, and you looked out for other people if you possibly could . . . two lessons that would stay with me for the rest of my life.

As bad as the Depression was, my family would have made it through all right if it hadn't been for an accident at the sawmill. My dad was working one day when a log rolled off and crushed his hips, legs, and chest, breaking a number of bones in the process. His injuries were extreme; it took him two years to recover, and even then he couldn't do the heavy work he'd devoted his life to doing.

That changed our lives in many, many ways. Not working meant no money; disability insurance didn't exist, at least not for regular people like us. We moved to Selma, Alabama, about twenty-five miles away. It was a larger place, where it was easier to get health care and other things now that the family was no longer self-sufficient. My brother Euel got a job driving a coal truck. I worked training horses and taking whatever odd piece of work I could find. I was only in high school, but I felt guilty—I wasn't pulling my own weight.

The only thing I could think to do was to drop out and get a job. So I did.

TREES, DOGS, AND LAUNDRY

The Depression went on and on. My father's youngest brother, Alvin, decided to give up the sawmill business. His new venture was delivering logs to a Selma lumber company. The money was better, or at least stable. I went to work for him. We harvested cypress along the river, often pulling trees out of standing water a foot deep, lopping off the runners and branches, and then loading them into a truck to be taken to the mill. Later, I took my uncle's place on a crew dredging the river so it would be deep enough for boats to safely travel.

I was fourteen, and was paid a dollar a day—a decent wage in the early 1930s in Alabama. To give you a comparison, a quart of milk in Birmingham, a far larger city not quite a hundred miles to the north, went for between eleven and fourteen cents. I was saving on rent, since I could live with my uncle, and soon I was able to buy some clothes to replace the ones I was wearing out and outgrowing.

By the time I was seventeen, I'd grown tired of working as a logger. I thought of going back to school. One of my cousins, Ralph Mims, invited me to live with him and his wife, Eunice, back in Clanton. I took up the offer, but only after he agreed to let me earn my way by doing odd jobs and much of the cooking, laundry, and cleaning. Every day I'd drive Eunice to the Isabella Agriculture School, where she taught math and I continued my high school education. I also played halfback on the football team.

A year of school was all I could manage. Money was so tight that I didn't have any left for clothes, and it was just too obvious to me that I needed to earn a living.

Ralph assisted me again, hooking me up with a local veterinarian who needed someone to help train his horses and generally help out. I went to live with the doctor and his family, and soon found myself appointed as the deputy veterinarian of Chilton County, a position that gave me a badge and a gun. My weapon of choice, though, was a hypodermic needle, which I used to give dogs vaccinations. The county took rabies pretty seriously; if your dog didn't have a tag proving he'd had his shots, the poor pup would be hauled off to the pound and kept there until the owner paid fifty cents for the vaccine.

I never had to use the gun, but the holster on my belt made my job a lot easier. Fifty cents was a lot in those days, and some of the farmers could be pretty ornery as well as protective, especially when it came to their dogs. I remember one of them coming to the door and swearing that he had no dog in the house—which was a hard position to maintain with all the barking behind him.

Our discussion ended when the dog shot out between his legs to confront me. The animal was vaccinated without further incident.

One of my favorite jobs for the doctor was training horses. It's a specialty that requires many things, chief among them patience. You can't force a horse to do something he doesn't want to do. You need to convince him that what you want is what he wants, and vice versa. You have to make him understand that if he trusts you, he'll succeed.

Getting an animal to trust you is a difficult but worthy task, one you can't fake. You have to be honest with yourself and the horse. You have to be dependable, willing to show your faith in him, and above all, you need to be patient and gentle.

Learn how to do it reliably, and you'll have the skills you need to lead men as well.

I trained horses, took odd jobs. I moved on to other temporary posts. But I didn't have a real calling. Worse, I didn't have a steady income.

Looked at objectively, and from a distance, I had a lot going for me. I was young and had a wide assortment of skills. I was good with animals. I knew my way around heavy equipment, both to work it and fix it. I was in good shape. I could cook. I could clean. No job was below me. I worked hard, and I didn't complain.

I did not have a high school diploma. That was a drawback. But it was also common at that time and place.

Even so, Alabama was still mired in the Depression, and there just weren't a lot of jobs to go around. And with older, more experienced men available, why take a chance on a kid not yet out of his teens?

As 1939 drew to a close, there was one large organization that was hiring, and that was not only willing to bring on a kid like me, but actually preferred someone of my age and skill set over older men.

The U.S. Army.

TWO

Echoes of War

YOU'RE IN THE ARMY NOW . . .

Nazi Germany's moves in 1938 to take over Austria and grab Czechoslovakia left no doubt war was coming, and that it would eventually threaten the United States. Even so, neither the American public nor our Congress seemed ready to address our military needs. The army had shrunk following World War I, and between that and the Great Depression, we were woefully undermanned and under-equipped.

Germany invaded Poland on September 1, 1939; Great Britain and France immediately declared war. Months later, Germany invaded Holland and Belgium, sweeping through the lowland countries on the way to France. By the end of the summer of 1940, Western Europe was a Nazi camp. Londoners ducked bombs every night, and spent every day expecting to be invaded.

While many Americans still believed the U.S. would escape the conflict, we were rebuilding and expanding our military just in case. My interest in the army coincided with the start of the buildup, just after President Roosevelt had declared a limited emergency and Congress had authorized appropriations to start funding modernization. The army, under Chief of Staff George Marshall, was already reorganizing, dramatically changing its combat structure to go with the new armaments and men it was adding.

I thought of none of that when I was deciding what to do. My goals were finding a job to help my family and to advance myself in life. Joining the army wasn't an easy decision, though. No one in my family, as far as I knew, had been in the military; we didn't have that tradition.

I talked to my Uncle Alvin about it. In many ways, he was the wisest of that generation in my family; he'd had more education than my father or their other brother, and he had a calm, logical way of working arguments out. We sat and talked a bit, and he agreed it was the best thing for me.

He also drove me up to Montgomery, Alabama, the nearest recruiting station as far as I knew. When we got there, I bid him farewell and went in to talk to the recruiting sergeant.

The station was at Maxwell Field, one of the army's preeminent air bases before World War II (and after, for that matter). Which probably explains why the recruiter tried to get me to join the U.S. Army Air Corps. (At the time, what became known as the "air force" was part of the regular army.) Of all the military branches, it had the most catching up to do. The air corps was getting new equipment, and training pilots and crews for what was still a very new phase in warfare, especially for the United States. I suspect my knack for fixing machinery might have made me an excellent candidate as an airplane mechanic, if they didn't want me as a pilot.

But I was a tough guy, or wanted to be. I told the recruiter I didn't want to go into the air corps; I wanted to be in a fighting unit.

"I want to be on the front lines," I told him.

Direct quote. What a stupid thing to say.

If the recruiter was insulted—or amused—he didn't show it. "Sure thing," he told me. "We'll get you into the First Division."

It was all the same to me, as long as I was going to be where the action was.

By then, it was pretty late in the afternoon, maybe even evening. The recruiter had me and another fellow go over to a hangar where we could bed down for the night. In the morning, we'd head over to the 1st Division at Fort Benning, Georgia, and begin our training.

I wasn't in that hangar too long before I got to thinking about what I'd committed to.

I was twenty years old, had never been able to finish high school, had worked all my life. I wasn't afraid of hard work, or even following orders, but did I really want to commit myself to army life?

What all did that entail? How dangerous was it, really?

And a million other questions, none of which I had the answer to.

So I changed my mind. I got out of the bunk, left the hangar, and started walking back to my uncle's house in Selma, twenty miles away.

He was surprised to see me come in the door. I guess that's an understatement.

"What are you doing here?" he asked. "I thought you joined the army."

"I need to talk about this," I told him.

And so we did. He made it clear I couldn't live with him anymore, and the truth was I'd already reached that decision myself. He had no work for me and I didn't want to be a freeloader.

I'd come to the conclusion that there wasn't any job around that I wanted or that would have me. So pretty much, the army was it. I just needed to talk it out.

In the morning, Uncle Alvin drove me back. I went and found the sergeant, who gave me a puzzled look and asked where I'd gone off to.

"It was a big decision and I wanted to be sure," I said. "I wasn't quite sure. Now I am."

He pointed to the bus for Fort Benning.

FIRST AND 16TH

I didn't know it, I didn't plan it, but I'd landed in one of the most capable and famous units of the regular U.S. Army, the 1st Division.

I'm partial, of course, but it's not just me saying it. Even today, the "Big Red One"—our division's nickname, which comes directly from the unit patch—is such a prestigious outfit that guys will continue to wear the patch on their "off" shoulder after they've been transferred—something that isn't allowed or even desired for most other regular army divisions. Past members of the 1st tend to look up others; I can't tell you how many times I've come out of the supermarket to find a fellow retiree waiting to chat me up by my car, which has a division plate on it.

Why is the 1st so famous?

Most simply because it was the "first"—in World War I, it was the first American division in France, the first posted to the front, the first to fire at the enemy, and the first called out for exemplary action. As a matter of fact, the 1st was literally the first division created in that war—we had no actual divisions until then.

You may have heard in history class that America's decision to join World War I on the side of the Allies dramatically changed the momentum of the war. What you may not have heard is that the 1st played a key role in the Battle of Cantigny in May 1918. The Germans were holding the village of Cantigny, France, protruding into the Allied line like a dagger that could strike the rest of the Allied army.

The 1st was tasked to take the village some fifty miles north of Paris. They not only seized it, but then managed to hold it after the Germans counterattacked. Realizing it was a strategic spot, the enemy threw everything it could at the division, which until then had no combat experience. More than one thousand men in the division were killed or wounded, but the Yanks (as our guys were called) held tight.

Not only was this an important strategic victory—it meant the Allies didn't have to worry about being attacked on their sides or flanks, the weakest points—but the 1st Division's courage and competence showed the French and British that we were up to par with the best fighters in the world.

That wasn't a simple operation, either. Not only did we have air support from the French—common now, rare in 1918—but the infantry made generous use of tanks and artillery, or combined arms, to use the military term. It was a modern fight in every sense of the word, a baptism by fire in the deadly tactics that would feature not only in World War I's endgame but in conflicts that would follow for years to come.

The 1st Division was also involved in Soissons, Saint-Mihiel, and the Meuse-Argonne, substantial battles later that year that helped seal Germany's doom. These were costly campaigns, and the division's reputation was earned with courage and a lot of blood. More than twenty thousand casualties were tallied by the time peace was declared that November. The men who led the division, Major Generals William Sibert, Robert L. Bullard, and Charles P. Summerall, are enshrined in army lore as some of the greatest individual commanders we have been fortunate to have.

Battle exploits, awards, and citations are tangible things. At least as important—much more important, I would say—is the spirit that a unit develops from an efficient and successful group. The proper term is *esprit de corps*. It can be hard to define—a culture of success, a reputation, a sense that you belong to an excellent organization that gets things—big things—done. The 1st had that spirit.

We also had a motto, adopted in World War I, that reminded us and others what we were all about: *No mission too difficult, no sacrifice too great—Duty First!*

Big tests lay ahead. The 1st would pass them all, and our esprit de corps and the legend of the division would only grow.

THE 16TH REGIMENT

Divisions are made up of smaller units, some with a special purpose, some that assist other units. In the American infantry, the bulk of the manpower is organized in regiments—large, mostly self-contained units that do most of the fighting. The U.S. Army had recently adopted a structure that assigned three regiments to each division. (In military terms, these are "triangular divisions," and they remain a common arrangement.)

Regiments as a general rule were commanded by a colonel, who in turn answered to the division's general. In theory and usually in practice, the regiments could be maneuvered or deployed in battle separately, augmented by smaller units or sometimes attached whole to other units depending on the need.

Each infantry regiment had three battalions, along with a headquarters company, artillery, an antitank company—and, most importantly as my service began, a medical detachment.

The battalions—and I'm making generalizations here—each had about nine hundred men, organized as companies. An infantry battalion would usually have a headquarters company, three rifle companies, and a weapons company. The rifle companies would list about 190 soldiers each.

The 1st Division had three regiments—the 16th, the 18th, and the 26th. I went over to the 16th, where I was to be trained as a medic.

If I had known next to nothing about the 1st Division, I was even more ignorant about the 16th Regiment. But if I had known—and had any say at all—it surely would have been the unit to ask for.

The 16th Regiment is one of the oldest units if not *the* oldest unit in the army. We trace our history back to the 11th Regiment—same unit, different name—which was formed May 4, 1861, less than a month after Fort Sumter was fired on to start the Civil War.

Mustered at Fort Independence, Massachusetts, the regiment was part of George McClellan's Army of the Potomac, taking part in the Virginia Peninsula Campaign and some of the most famous and bloody fights of the war, including Second Bull Run, Antietam, and Gettysburg. Its soldiers were at Appomattox to disarm the rebels when Robert E. Lee surrendered.

Have you heard about Teddy Roosevelt and his famous charge up San Juan Hill during the Spanish–American War? The 16th helped take the hill. Later it was sent to the Philippines and saw action there. General John "Black Jack" Pershing led the 16th and sister units in action against Mexican bandits and Pancho Villa. In World War I, the 16th had the honor of parading through Paris, announcing the arrival of American combat troops—or as Black Jack Pershing put it with the words "Lafayette, we are here!" letting France know we were repaying the debt incurred when the French and the Marquis de Lafayette helped us win our independence.

The 16th was on the receiving end of the first bullets fired by Germans at American troops at Bathelémont; as part of the 1st Division, the regiment was involved in the fight for Cantigny and other critical battles as the Allies pushed the Germans back and pressed on to victory.

It was a great unit. Something to be very proud of, as I would soon learn.

FROM DOGS TO SOLDIERS

When the recruiter asked me what sorts of jobs I'd had, I led with my stint as veterinary assistant. That apparently impressed him and the rest of the army, because I was assigned to the 2nd Battalion's medical detachment as a medic.

I guess they figured if a man can take care of dogs, soldiers would be a cinch.

Truthfully, there were similarities beyond the fact that GIs at the time were often called "dogfaces." The most important thing I'd done as a veterinarian's assistant was give rabies shots; plunging a needle of morphine (or whatever else was called for) in a soldier's arm was relatively easy after that. I can't remember a soldier trying to bite me, which is more than I can say for the dogs. And if my qualifications on paper didn't amount to much more than that, they were actually better than those of most of the recruits who found themselves in the medical corps.

Doctors were officers, and generally in charge, at least in name—as always, a lot of the real work and authority was done by the NCOs, or noncommissioned officers. Each battalion's medical detachment had about thirty men total, including a doctor and a staff sergeant in charge. All had training in first aid, though this was fairly basic care. In the field, the medics would be divided among three different jobs. Some would work directly with the companies—"company men"—caring for anyone hurt or otherwise in need of care. Some would man a battalion first-aid station close to the front line. The last group—the stretcher bearers—worked in between.

As a raw recruit, a lot of my time after I learned how to march and the basics of getting along in the army was spent doing non-medical things. Medics at Benning didn't do advanced infantry training, and we were used more for driving vehicles than applying bandages—which made sense, I suppose, since not only were there few injuries to deal with, but we weren't very deeply trained in medicine at that point.

Medicine, and especially battlefield medicine, was very different in those days, nowhere near like it is today, where helicopters can whisk injured GIs directly off the battlefield and the care they receive on the way to a hospital can equal what they would get at a trauma center. I've met Special Forces medics who are practically doctors.

Back then, our job was much simpler: stop bleeding, usually with a tourniquet; clean wounds; inject the wounded man with morphine to ease his pain. (A tourniquet would cut off blood to a limb above a traumatic wound, preventing the patient from bleeding out. Its use in combat dates back at least to Alexander the Great. There are complications from using tourniquets, however, and the practice has greatly declined since my war.)

The soldier would then be taken—usually by two stretcher bearers, who were also medics—to the battalion first-aid station near the front lines. There he would be treated by other medics and, if warranted, the doctor in charge of the unit.

If the wounds were light, the soldier might simply walk back to his unit. Serious cases would be evacuated to a larger station—the regimental aid station, usually close to the regimental headquarters a couple of miles back. There they would receive further treatment before being shipped on to a collection point and finally a hospital.

The division had only just been assigned to Fort Benning, and our area, Harmony Church, was more pasture than military camp.

Muddy pasture, I should say. We were out in tents, four men in each, and one of our first jobs was to build wooden platforms to keep our feet dry, or almost dry, when it rained. We built sidewalks as well, and soon the place began to actually look like something.

The army was growing quickly, and as guys even fresher than

me joined the unit, I realized I already knew quite a lot. Not just about bandaging or setting compound fractures. We got many kids from cities like New York or Philadelphia who had spent all their time in the city, and didn't know much about machines or how to fix them—a handy skill when your Jeep or ambulance breaks down.

I was always ready to learn something new, and to take on another job. People tell you never to volunteer when you're in the army; it will only lead to trouble. Fortunately, I hadn't heard that advice. The officers and NCOs in charge appreciated the fact that I was willing to do whatever they asked, taking on jobs that I'd never had before. They gave me more and more as time went on. Not only didn't I mind, I liked it.

BROTHERS IN ARMS

I'd been in the army for only a few weeks when another new recruit showed up at our camp—my brother Euel. He and I had talked about his joining as well, so it wasn't all that surprising to find him in khakis. He'd joined a different unit, also training at Benning—the cavalry, which was still part of the army in 1940.

My brother had asked to be assigned there because he loved horses, and I imagine he had some romantic ideas about what that might involve in the army. But the reality was not very pleasant. As he put it to me after a few weeks, "All I'm doing is shoveling shit in the stables."

"Join the medics," I told him.

I guess I made the job sound attractive, or maybe anything sounded good after a few weeks of handling manure. He quickly agreed. We went and talked to my commanding officer, and Euel joined the medical department and came over to the 16th Infantry. He'd prove to be just as good a medic and leader as I was, as time would tell.

Euel never really liked his first name, and for some reason I either never knew or have long since forgotten, he acquired a new nickname soon after joining: Bill. It was our father's name, which may explain why he favored it. In any event, people started calling him that and it stuck for the rest of his life—just as I will for the rest of the book.

People have asked me if having my brother in the same unit made me worry about him, especially later on during combat.

I don't think I ever worried about him. I was *concerned,* maybe.

That's a different thing. When you're worried, you're constantly thinking of someone or something.

When you're concerned, it's not in your mind constantly.

We were both still kids in some ways. Maybe I thought of myself as indestructible, and him even more so, because he was my *older* brother. Nothing could happen to either of us.

You have a lot of optimism when you're young.

Our training became more complicated as the weeks and months passed. When the division went on maneuvers—the military term for practicing combat or combat-related procedures—the medics would set up an aid station. If there was a mock battle, we would treat mock wounds, as well as the occasional sprained ankle or whatnot.

A typical aid station would be set up in a large tent marked by Red Cross flags. An operating table—a stretcher perched on our two large medical chests—would be set up near the middle of the tent, off-center enough to allow for an area at the side where the walking wounded could be treated. We had a station where we'd clean and disinfect instruments with alcohol. The rest of the space would be taken up with supplies.

Things were more ad hoc during amphibious landings; you weren't going to set up a tent in the water, or give the enemy an obvious target. And the wounded were always the first priority.

Besides learning how to use things like tourniquets, properly clean wounds, and set broken bones, we were taught how to give blood transfusions and use plasma on the battlefield, along with more complicated procedures for assisting the doctor. Though plasma would turn out to be a lifesaver, there weren't too many opportunities to use it at the forward aid stations where I would work during the war.

I was sent for medical training to Fitzsimons General Hospital in Denver, Colorado. Fitzsimons and the U.S. Army played an important role in battling tuberculosis, which was a devastating disease at the time. TB, or consumption as it was known in earlier times, is an infectious disease that had plagued the army in World War I. Fluoroscopy, an early form of X-rays, and other screening helped identify the disease in its early stages, allowing patients to be isolated and treated before infecting others. At the hospital I was taught how to remove fluid from lungs, placing a needle to drain the organ and then forcing the liquid out.

We also trained to assist during surgery, and learned how to set bones by working with animal bones. Here my experience with the veterinarian and working on the farm really did pay off.

Besides becoming familiar with how to treat injuries and wounds, we developed the professional attitude you need to treat the injured in battle. You have to be able to distance yourself from the person you're helping; if you think too much about the patient as a person, a fellow soldier, or even a friend, your feelings can choke out your ability to help him. Not a good thing.

I spent about six months at the hospital, learning and working with real patients. When I got back to my unit, I shared what I'd learned with the others. Not to brag, but Command was already looking at me to teach and lead. It was the sort of thing I liked to do. I guess I had a talent for it, even though I'd never had a chance to develop it before.

SUGAR "RAY"

Our training and that of the entire army ramped up; we took part in the Louisiana Maneuvers, a massive exercise that, looking back, illustrates exactly how depleted and out of date our army really was: soldiers marched with sticks, and trucks stood in for armored vehicles by posting the word "tank" on a placard on the side. We were moving in the right direction, but with fits and starts.

The army wasn't all training and hard work, though. We had our share of recreation. In my case, that meant boxing.

The sport was very popular at the time. Units would pit their own champions against each other, and competitions between battalions and divisions were a hit with the troops. I imagine there was a bit of bragging among the brass as well.

I'd been fairly athletic as a kid, and I could hold my own with my fists. Let me say I weighed somewhere between 145 and 170 pounds at the time—a welterweight at the lower number, a middleweight at the higher range. I'm not saying I was a great fighter, but I did well enough in the ring for some of the guys to compare me to a very famous boxer who dominated in those weight classes—Sugar Ray.

For you younger readers, that's Sugar Ray Robinson, an African American who boxed in the 1940s, and whom a lot of boxing historians still regard as one of the best pound-for-pound boxers of all time. More recently, another outstanding welterweight earned the nickname—and accolades—Sugar Ray Leonard.

I was nowhere near as good as them, of course. But everybody gets a nickname in the army. Mine became Sugar Ray. "Sugar" was quickly dropped, and from then on, like my brother, I had a name that would follow me through life.

Boxing was recreation, but we took it seriously. We trained hard. In one case, we journeyed to New York City and Stillman's Gym, the city's famous university for world-class professors of boxing science—I mean professional boxers. We trained for three days before heading to upstate New York to fight matches in Plattsburgh with other members of the division.

Stillman's may have been a pugilist's palace, but it was also a gym, a *real* gym, which meant it smelled of sweat and probably blood. You didn't want to eat there, off the floor or anywhere else. But it was a thrill to be in the place where the famous names of boxing past and present had learned their craft.

The bout I remember best pitted me against a guy from another unit in the division who turned out to have been a pro and held a championship belt in his life before the military. He didn't knock me out, but he got me pretty good. I may still have the bruises to prove it.

And then there was Smokey.

Smokey was a friend of mine in our unit, an Italian American who, for all the world, looked like a character right out of the funny pages—what we called the newspaper comics back then. He was short and fought as a flyweight; he was probably punching up a weight class or two even so. He always had a sawed-off cigar in his mouth and reminded us of the sparring partner of a famous cartoon fighter, Joe Palooka. Who of course was named Smokey, which is where his nickname came from.

Tough little guy.

Smokey and Palooka, by the way, would follow us to war. But while there were premonitions, we didn't know what lay ahead.

I'm not sure what we would have done if we did. Despite the buildup of the army, the country wasn't ready for real war. People didn't think about our position in the world anywhere near the way we do now. The Atlantic and Pacific Oceans seemed like immense barriers, and even folks with relatives in Europe and Asia tended to think the U.S. was immune to the problems over there. Europe was the "old world." We were the new.

Maybe so, but that wasn't going to protect us too much longer. We were about to find out that you can't hide behind an ocean when the rest of the world is at war.

I know from our perspective now, the war should have seemed inevitable. By the fall of 1940, London was being bombed nightly. The attacks were gruesome. A reporter I later met and admired, Ernie Pyle, described one of the attacks with words that sound almost like poetry, yet are awful in their grim reality:

London stabbed with great fires, shaken by explosions, its dark regions along the Thames sparkling with the pin points of white-hot bombs, all of it roofed over with a ceiling of pink that held bursting shells, balloons, flares and the grind of vicious engines.

How could anyone imagine any corner of the world would be safe after that?

ROMANCE, THEN LOVE

There were plenty of other things to focus on, at least for me.

Like love and romance.

Soldiers were popular with the girls, and wherever we were stationed, the USO or some similar organization would put on dances and the like. We were moved up to Fort Jay on Governors Island in New York City, and that was a gold mine for dating. Girls would come over and we'd meet for the dance at the USO building. Unfortunately, they had a strict rule—once inside, no girl could leave without a chaperone. And there were MPs at the door to enforce it.

That cut down on the "necking" for sure. But you could get the girl's phone number and maybe a date for later on.

Manhattan was a ferry ride away. We'd go over to Times Square and have our pick of the ladies. And the entertainers and big shows were pretty good to servicemen. I remember the time we all got tickets to see an up-and-coming singer. The tickets were free, on the condition that we bring a date. The young singer's promoters wanted the girls to sit in the front rows and scream.

Really. Someone would hold up a sign on the stage telling them to do just that. The show was being broadcast live on the radio, so to the audience the screams seemed spontaneous. So we had hype as well as sex back then.

The singer's name was Frank Sinatra. I guess those promoters knew what they were doing.

But it was actually in Denver, when I was sent for training at Fitzsimons, that Cupid's arrow first found its bull's-eye in my heart.

We had a day off, and three or four of us went out to a riding stable where we could take out some horses and picnic. We were near the barn when a group of girls came over and started chatting. They were pretty and easy to talk to. I hit it off with a girl whose name was Helen.

"What do you do?" I asked.

"I'm a singer."

"That's nice. Maybe I can hear you sing sometime."

"How about tonight?"

She told me she was singing at a club nearby that evening. Naturally, my friends and I had to go to the club. I didn't quite realize that the young woman I'd taken a fancy to was a celebrity, the lead female singer in the Jimmy Dorsey Orchestra.

Helen was Helen O'Connell. Not that the name meant much to me when we first met. Nor did her fame. What mattered was that she was fun to be with, as good a listener as she was a singer. Smart and vivacious, pretty—I could go on for quite a while with all the adjectives and still not describe her quite right.

Is it any surprise that "Green Eyes," a song she made popular with Bob Eberly and the Dorsey band, is still one of my favorites?

Between her sets that first night, Helen came and sat with us.

Over the next few days, while we were both in Denver, we spent as much time with each other as we could. If it wasn't love—and I'm not saying it was or wasn't—it was certainly an infatuation on both our parts.

It continued when I went back east. The Jimmy Dorsey Orchestra is not as well known today as the outfit of a similar name led by Jimmy's younger brother Tommy Dorsey, but in its day it was headline material, with a string of hits from 1939 well into the war years. The musicians went all over the country, performing for full houses. Jimmy Dorsey played clarinet and saxophone, and a number of jazz greats credit him with influencing their style. The band's swing-style jazz often had a strong, very danceable beat, and duets between Helen and Eberly were both innovative and extremely popular.

Helen's travels took her not only to New York and Massachusetts, where my unit was later assigned, but the South. We were able to see each other several times. But time and distance work on love affairs like acid on metal. By the fall of 1941, our passions had cooled, or maybe the reality of our different directions had set in.

But I had set my sights on another young woman, one just as pretty: Estelle Saunders.

ESTELLE

We were on maneuvers in North Carolina, camping amid the peach groves in the rolling hillsides west of Fort Bragg near Biscoe. Sunday came along, and I headed over to a local church. It had a small, friendly congregation, and when the service was over the deacon invited a group of us to his house for Sunday dinner, the big midday meal.

It turned out the deacon had two daughters—Estelle, about my age, and Becky, a few years younger. Estelle was a knockout. That wasn't just my opinion; she'd been a runner-up in a Miss North Carolina contest not too long before.

There was a lot more to her than looks. She had a friendly personality, a forgiving nature, and a lot of patience. I found her very easy to talk to. Something sparked between us almost immediately—love, whatever that is.

When the time came for us to get back to camp, I asked her if she minded if I called on her.

She did not mind. On the contrary, she rather liked the idea.

I went back two days later. We quickly became an item. I saw her as much as I could while we were camped nearby, and then kept in touch as much as possible over the next weeks and months. It was a quick, intense courtship, but not without its hurdles. I think it's fair to say that Estelle's mom was not thrilled with the idea of a soldier seeing her daughter. She assigned Becky as our chaperone anytime we wanted to go out. Efforts to get around this were unsuccessful; the single time we managed to sneak off alone to a dance place—more like a small roadhouse out on a country road—Mrs. Saunders found out and sent her husband to fetch us home.

Worse, she arranged to sit in her front room every time Estelle and I sat together—or almost together—on the porch at the front of the house. Hard to steal a kiss with Mom glaring through the window at you.

As far as I can remember, I managed it exactly once.

As sweet as that kiss was, I wanted more. With my unit now

assigned to Fort Devens in Massachusetts, the prospect of seeing Estelle on a regular basis was pretty dim. Unless, of course, we married.

Now, here I was, in the army, no job beyond that. No money to speak of, beyond Uncle Sam's paycheck. What right did I have asking someone to marry me?

None. But I did love her. And I was sure I would spend the rest of my life with her.

I guess she cared a lot about me, too, because she looked past the negatives and said yes when I asked.

The deacon advised against it—in fairness, we hadn't known each other all that long, and I was making the not-too-princely sum of twenty-one dollars a month in the army, a poor wage even then. But Estelle persuaded him to approve. I don't know if it was me he valued, though I like to think that I cut a good figure as a prospective son-in-law: I didn't smoke, I rarely drank alcohol, and in fact to that point in my life had never been drunk. I went to church, and I was advancing through the enlisted ranks in the army. I was from the rural South and had worked hard all my life.

Then again, Estelle could be very persuasive. In any event, her father not only gave his blessing, but promised to bring her mom around as well.

We started talking about wedding dates.

Then, right around the time we were planning to get married, our world changed dramatically: the Japanese bombed Pearl Harbor.

THREE

Infamy and England

"REAL WAR"

Your mind plays tricks when you look back. Things that should be sharp and crisp blur. Odd events, people you barely knew and places you rarely visited, suddenly become sharp.

It's not just age, though that's there. It's just how the mind works.

So when I say that I'm not exactly sure of where I was when I heard the first news of the Japanese attack, or the circumstances that brought me there, I hope you'll understand.

I think I was in Selma, possibly for my grandmother's funeral. I know Estelle and I were planning on getting married right around that time. It was undoubtedly a normal, quiet Sunday in every respect, right up until mid-afternoon. The first we heard was probably from neighbors, telling us to turn on the radio and hear reports similar to this one, which came from Honolulu:

Hello, NBC. Hello, NBC. This is KTU in Honolulu, Hawaii. I am speaking from the roof of the Advertiser Publishing Company Building. We have witnessed this morning the . . . full battle of Pearl Harbor and the severe bombing of Pearl Harbor by enemy planes, undoubtedly Japanese. The city of Honolulu has also been attacked and considerable damage done . . . It is no joke. It is a real war . . .

My brother was with me in Selma, and our first thought was that we would be immediately called back to the regiment. As best I can remember, he called headquarters that afternoon and managed to talk to someone in charge, who said no orders had been given to return. Estelle and I went ahead with our plans, and married on January 8, 1942. There was no honeymoon. I headed back to Massachusetts and Fort Devens right after the wedding.

1942: WORLD AT WAR

The winter of 1941–42 was the low point of the war, not only for the United States, but for the Allies in general. America declared war on Japan December 8, 1941. Germany declared war on the U.S. a few days later, and of course we returned the favor.

Japan was on the move, taking Singapore in mid-February and the Philippines by May. Singapore was British; the Philippines, an American commonwealth pending a transition to full independence. The American army and Philippine soldiers fought an extended campaign against the Japanese, who though outnumbered steadily increased pressure and tightened the noose around the defenders over time. Douglas MacArthur's famous "I Shall Return" speech sounds thrilling now, but at the time it was only a bunch of very brave words, and very possibly an empty prediction.

Not that we soldiers felt that way. We were eager to fight, angry that we'd been attacked, and sure that we could beat anyone. Maybe too sure.

In Europe, Adolf Hitler had launched an invasion of the Soviet Union in June 1941. His armies had stopped just short of Moscow. The long Russian winter would take its toll on the Nazis, but that wasn't clear in the first half of 1942. The German army went back on the offensive once the weather cleared. The Russians suffered immense casualties, far greater than the U.S. or any of its allies eventually would. The battle for Stalingrad, now seen as the turning point in the war on the eastern front, was far in the future.

Half of France was occupied by the Germans; the other half was a puppet state. Jews were being rounded up and taken to be slaughtered. The Germans were still bombing England.

They were also waging war in northern Africa, working with their Italian allies. Since 1940, they had fought back and forth with the British in the desert and high hills. With the arrival of German general Erwin Rommel and the relatively well-equipped units that would form the Afrika Korps in early 1941, the Germans began building momentum. In January 1942, Rommel shocked the British by launching an offensive just at the point where they thought his forces were depleted. Helped by newly arrived American tanks and other equipment, the British avoided total catastrophe, but only just. In June, they would lose Tobruk and 351,000 men would enter the rolls as POWs. This defeat, known as the Battle of Gazala, ranks as one of Great Britain's worst in all its history. It would turn out to be Rommel's greatest victory and his high-water mark.

With war waging on both sides of America, President Roosevelt, Army Chief of Staff General George C. Marshall, and our other leaders debated where to put our country's priorities. While they decided the U.S. would fight a two-front war, taking on Japan and Germany at the same time, Europe became the main priority for a number of reasons, not least of which was the fact that Hitler seemed very close to total victory there.

Stop him now, while Great Britain was still in the fight and could be used as a base and staging area, or never stop him.

A LAST NIGHT WITH MY WIFE

I, of course, had nothing to do with those sorts of decisions; as a T-4 and then a T-3—"Technician 4" and "Technician 3," ranks roughly equivalent to and usually called sergeant—I worried about my job and my unit, and little else. At the same time, there were plenty of rumors about which way we were heading. Mostly, they predicted that we'd ship out to Great Britain.

We kept training. I wangled my way into special rifle training, qualifying as a marksman and earning a badge. Ordinarily, medics didn't carry weapons, not even pistols; our job in combat was to help the wounded, and according to the Geneva Conventions we were not supposed to fight or be fired upon. In combat, our helmets would have large red crosses; we would have armbands with the same very visible insignia.

I took the course anyway. It's possible I was the only medic who did that, at least in the 16th. Since I'd hunted from the time I was a boy, the course wasn't all that difficult; I imagine a lot of guys who'd grown up in farm country found it a breeze, especially when it came to firing the M1 Garand. The rifle was a revolutionary battle weapon, a semiautomatic firing a .30-caliber bullet from an eight-round clip. You could fire as quickly as you pulled the trigger, a great advantage in combat over the earlier bolt-action Springfield and the rifles used by other armies at the beginning of the war. The recoil was light, and the iron sights on the gun allowed an experienced rifleman to adjust for conditions with good results out to a few hundred yards.

I earned a Marksman Badge with Rifle bar when I was done. Being good with a gun was a fine skill to have, no matter what your actual job was. If nothing else, it gave you the confidence that you could defend yourself if things went really bad.

At some point in late spring, around the beginning or middle of June, we were sent to a camp in Indiantown Gap in Pennsylvania and put under quarantine. The time had come to ship out. We were headed for Great Britain, and I was selected to lead an advance party of medics.

As excited as I was to do the job, I had one reservation: I knew we'd be gone for a long time, and I wanted to see my wife before I left.

Impossible.

But nothing is truly impossible, not in the army. I had a talk with my commanding officer.

He was sympathetic but made no promises. The next day or thereabouts, he told me that I would be allowed to meet my wife in town for one last good-bye.

"Don't breathe a word of it to anyone else," he ordered.

"Great," I said. "But if I can't leave the base to make the arrangements, where are we supposed to go?"

He frowned. I guess he hadn't gotten quite that far in solving the problem. But he soon took care of it. I got word to my wife to meet me in town, and on the appointed day an MP showed up at my barracks and took me to an apartment that had been set aside for us.

We thoroughly enjoyed our hours together. So much so, in fact, that my son Arthur was conceived. Not that either of us knew it for certain at the time.

ENGLAND AND ALL

We left for England on June 30, 1942, in *The Duchess of Bedford,* an impressive two-smokestack cruise ship that had been converted for use as a troop carrier by the British. Before the war, the Canadian-owned steamship had gained a reputation not only for luxury but for rolling in heavy seas, earning the nickname "Drunken Duchess."

The 20,000-ton liner was one of eight troop and supply ships that sailed with sixteen escorts in the middle of the night from New York Harbor for Liverpool. There were between twenty-five and thirty guys with me, assigned to various tasks. Two or three were medics, who were to help me scout out the facilities and get ourselves ready for the full regiment, which was due a few weeks later.

A once-elaborate dining hall belowdecks had been stripped and turned into a combination mess hall and dormitory outfitted with swinging hammocks. The fixed tables were meant to fit sixteen men apiece for meals; it was a tight squeeze.

Aside from the waves and crowded conditions, it was a quiet passage for the first few days. Then came the night we heard shouts on the deck. A submarine had fired a torpedo at the convoy.

Our escorts scrambled to the attack. Up top, I watched a destroyer dart out through the shadows and begin launching depth charges. The water rumbled below.

Did we get him?

I had no way to tell. The *Bedford* and the rest of the convoy continued eastward; there was no stopping for any reason, and lingering under these circumstances would have been foolhardy, as German U-boats often operated in packs. I heard later that the U-boat the convoy attacked had been sunk. But I also heard that it had been captured.

Alas, though we've looked, I've never seen any hard proof that any of that happened. There are various rumors that the *Bedford* herself sank a submarine in 1942; probably best to take them all with a grain of salt.

The submarine chase was our only excitement before we reached Liverpool on July 10. From there, we boarded a train for Tidworth Barracks near Salisbury, England.

A marching band struck up a tune as we stepped off the train. We attempted to march in time to their beat as they escorted us to the installation. But their tempo was far faster than what we were accustomed to, and I finally told the guys to just walk.

We saw there would be other things to get used to when we headed over to the mess hall. I'd no sooner sat down than one of the members of the women's auxiliary that had come to help us prepare the barracks raced over to my table.

"You can't sit here," she said sternly.

I asked why.

She pointed to the stripes on my shoulder.

"You're an officer. You need to go to the proper dining room."

At some point I'd been told that, to the British, noncommissioned officers or sergeants were officers, and were treated (and expected to act) as such.

It seemed odd to leave my guys, most of whom I'd known for over a year. But there was no arguing with her; she could have intimidated Winston Churchill himself. I got up and headed to the adjoining room.

I'm pretty sure we had the same food in both places; it was only the company that was different. Dinner that night was sausage and cabbage. The next meal, things switched up—we had cabbage and sausage.

It went on like that for the rest of the time we were at Tidworth. The British had been at war for several years now, and food was not in great supply. I'm sure what they were presenting us was the best they had, very likely better than what their own soldiers were eating and certainly better than what many civilians could muster. I began to appreciate the sacrifices the British had made since the war began.

LIVE FIRE

Our barracks were old two-story buildings dating from the turn of the century. The layout was a lot like the barracks we'd had in the States, with small offices on each floor and large open areas. The women who'd been assigned to assist us did a great job cleaning and helping us get new straw—yes, straw—into the bunks to provide a cushion for our bedding.

They were also the first women most of us had seen for quite some time. Which explained why on the first morning after we got there, when I asked the man I'd put in charge of bed check—standard procedure at night, kind of a roll call to make sure everyone was accounted for—he explained to me that bed check had been very easy: no one was there.

Apparently all the guys had been getting acquainted with our hosts.

Not a problem, I decided, as long as everyone was present now and there had been no bad behavior.

After things were set up, we left to join the rest of our unit, which was arriving in Scotland for training before going to Tidworth. We practiced amphibious landings and started doing live-fire drills—people were shooting at us, or in our general direction. It was the first time we'd had something more than firecrackers simulating battle conditions.

You practice and practice and practice, trying to make everything automatic.

The other thing we had a lot of in Scotland was rain. In my imagination looking back, it rains every day. I'm sure it didn't, but it seemed that way even then.

There were so many maneuvers, both in Scotland and the rest of Great Britain, that they blur together now. I do remember one time in Ireland, or rather sailing there. We were all on a ship, and the weather was so nice that a lot of us went up on deck and basked in the sun without our shirts—against orders.

I almost got court-martialed for that. Because between the salt water and the sun, I, like a lot of other guys, ended up with sunburn and blisters. I'm pretty sure that if one of the doctors hadn't intervened on my behalf, I would have gotten in serious trouble; there was no hiding the fact that I'd exposed myself to the sun.

A small little thing for a civilian, maybe—a bad day at the beach. But imagine the effect if we'd been going into combat.

With all the training we were doing, there wasn't a tremendous amount of free time. We did on occasion venture to a local pub. I remember one where the old guys were over at the far end, playing darts and drinking a light ale. I went over and talked to them a bit. When it was time for a fresh round of beer, they would go over to the fireplace, take an iron poker from the side, and stick it in the fire. When it was red hot, they'd plunge the poker into their drink with a sizzle.

They claimed it made it stronger. I couldn't say.

Truth is, I'd never been much of a drinker, and if I'd even tasted anything stronger than beer to that point it would surprise me. I'd never seen my father or one of my uncles drunk, at least that I can remember. In fact, I can't even picture them wanting a drink. That's what growing up in the Bible Belt does.

Now, it might have been that there were a bunch of drunks in every house I stopped in, all hiding their booze under the kitchen sink. But I didn't see it.

Getting back to England, we did take one sightseeing trip, or at least it was something like a sightseeing trip. The battalion marched out to Stonehenge at some point, and while there, got a brief, informal lecture on the purpose of the ancient stones.

Not exactly a package tour, but it was a bit different from the usual maneuvers. And I still admire the ingenuity of the people who built it.

My responsibilities had increased to the point where I was doing the work of the detachment's staff sergeant, the highest-ranking enlisted man in the unit. While each detachment was led by a doctor, who was generally a lieutenant, as a rule they left most of the day-to-day operations and direct supervision of the men to the ranking NCO. The doctors were highly trained in medicine, but most if not all had been in the army for only a very short time. And I guess from their point of view, they had more important things to do than watching over some thirty-odd medics.

We did have a staff sergeant, who was supposed to be running things. To put things in the kindest way, he wasn't much good at his job. But honestly, I didn't mind doing the work. My attitude was simply to get things done. If that meant taking responsibility for the others in the unit, what was the problem?

My brother Bill, in the meantime, had been promoted to staff sergeant, heading the medics over at 3rd Battalion. He had the rank—and the pay.

He was also, like me, a married man.

He'd broken his arm back at Devens. Sent to the hospital, his recovery seemed to take forever, but I eventually learned why. The nurse taking care of him was an intelligent and beautiful young lieutenant, Una Curran. By the time he was discharged, and probably well before that, they had fallen in love. Una resigned her commission to marry him.

BIGGER AND (MAYBE) BETTER

The army and navy were growing, adding men and equipment as quickly as they could. The draft had been instituted in 1941, and by Pearl Harbor, there were 1.6 million men (and a few women) wearing army uniforms.

Sounds like a lot. But consider this: By April 1945, there would be over three million American soldiers in Europe alone. The U.S. Army totaled five million soldiers—3.5 percent of the entire population. Account for the fact that women, children, and older men were largely excluded from service, and you get the picture—the odds were that you knew someone who wore khaki fatigues to work every day. It may well have been your brother, your dad, your uncle—or you.

(Today's numbers: the army has less than half a million men and women, about .3 percent of the population.)

My job as a leader—with or without the staff sergeant stripe—was to build our unit into a team that could work together under the worst possible conditions. To do that, the men had to know and trust each other. They also had to be put into positions where they could succeed.

Living with them, especially later on when we got into combat, I came to see them as family. You know, you live with guys and you get to be very close to them. You're concerned about them. You know that the slightest bit of a problem they may have could become serious when you go into action. So you get to know your people.

I spent as much time as I could getting to know every man, asking about their families, their backgrounds, finding out what they liked and didn't, watching them in the drills and later in the field. I have to say that most of them were top-notch. A few might be better adapted for one sort of role or another, but pretty much without exception they strove to do their best.

Aside from the staff sergeant.

The only other exception had been transferred out while we were at Devens. He wasn't a bad person, really, but he was always finding his way into some minor trouble or coming up short in an assignment. To put it in contemporary military terms, he wasn't squared away.

I am being kind.

The fellow spent a lot of time on KP duty—"kitchen police," where he'd have to do things like peel endless sacks of potatoes as a disciplinary action. Eventually, the commander saw fit to have him transferred out of the unit.

I heard later he eventually became a major. I'd like to believe that means he straightened himself out, but it may be a more accurate assessment of how hard up the army was for officers.

Not to mention ample justification for the ordinary enlisted man's view of the officer class, exceptions duly noted.

As summer turned to fall in 1942, the regiment started receiving new equipment. That was the surest sign we were going into combat, though at that point we had no idea where it might be.

In a sense, it didn't matter. Most of us were eager to get in the fight. We'd trained for years, some of us, and we were eager to put our skills to use. While there may have been a few men with reservations, most of us thought we were ready.

More than ready.

I certainly felt that way. I may even have been a bit cocky about it. I felt I knew what to expect, and was sure I'd do well.

But there are things that you can never really know until you experience them. And as we were all about to find out, war is one of them.

FOUR

Lighting the Torch

LIGHTS

The glow surprised me. It seemed unreal, to come from a time and place that had passed out of memory. I stood on the deck and stared, stunned for a moment.

We were off the coast of Spain. The lights were on because Spain was not at war; it was one of the few places in Europe that had not only remained neutral but had managed to escape combat or occupation, at least partly because its dictator and his government had close ties to Germany and Hitler. It was still recovering from a bloody civil war, and divisions and hatred no doubt still split the population, but compared to the open violence and deprivation common in the rest of Europe, it was a relative paradise.

With its lights on.

Of all the things I'd seen since joining the army, the last thing I would have expected to shock me would be the lights of a city. But they did. We'd lived in blackout conditions, not only in Great

Britain but the U.S., for so long that electric lights at night were an alien thing.

Under other circumstances, maybe they would have been a treat. But we were heading to real war, and the reality of that had begun to slowly sink in since I'd boarded the *Bedford* a few days before. It was mid-October 1942, and not counting a few skirmishes, most notably a Ranger attack on the northern French port city of Dieppe in August, the U.S. was about to spill its first blood in the European theater's ground war by invading Africa.

Once again I was part of the battalion's advance team, though this was a very different mission; instead of brass bands, we expected to be greeted by bullets from French Vichy soldiers under the direction of their German masters.

We—I, especially—were still cocky and edging for a fight.

I'd set up our aid station on the upper deck, which gave me more room to work and somewhat better conditions than most of the soldiers crammed below. I shared a cabin, small but comparatively comfortable, with one of our company aid men.

Captain Samuel Morchan, our doctor, headed our contingent. Just thirty years old when I met him, the doctor would prove a steady and calm presence, not only while treating wounds but under fire as well. That wasn't just my opinion—the army ended up pinning Silver and Bronze Stars on him. He'd joined the medical department in 1940 after a few years of practice and studying in Switzerland as well as the States.

The upper echelons had feared submarine attacks or even action from Spain as we passed through the Strait of Gibraltar. But the passage was quiet, and the entire armada of over two hundred ships made it offshore of Africa without incident. If we were detected, neither the Germans nor the French made any effort to stop us.

The passengers who'd sailed on the ship in its earlier incarnation would have leapt overboard if they'd had to eat some of the food being served in the cramped galleys below. Stories later cited not only the endless servings of chewy mutton, but also the liberal sprinkling of weevils and other bugs in the flour and bread. Soldiers broke out their emergency candy bars in self-defense.

My position as medic and my office above came with some perks. I may not have gotten any better chow, but I didn't have to snake through the seemingly endless lines. The galley—the navy's word for kitchen—provided food whenever we wanted.

Sorry to say, the cabbage and mutton didn't grow on me.

The candy bars were standard equipment—D-Rations. They weren't candy; the idea was that they would perk you up when you needed a boost. In fact, they'd been formulated to be a little bitter to prevent soldiers from treating them as candy and snacking on them rather than saving them for when they were truly needed. That didn't work all that well. They were still sweeter than just about anything else a guy had eaten for weeks.

C-Rations were more substantial. These were canned meals that were supposed to substitute for hot food while you were in the field. They came in three varieties: meat and beans, hash, and stew; others were eventually added, including my "favorite," franks—or whatever that meat was—and beans. That was probably the favorite of quartermasters as well, since those always seemed in short supply at the front.

These main courses were supplemented by crackers, juice powder, instant coffee, and salt and pepper. Early on, there were "energy tablets"—dextrose or sugar—as a dessert; this got switched to candy and then cookies as the war went on.

We also got halazone tablets to clean drinking water, chewing gum, and of course cigarettes as add-ons. The gum and the smokes were more popular than the canned food.

ORDER OF BATTLE

The area of northern Africa we were heading toward had been in Axis hands since the fall of France in 1940, though not directly held by Germany.

The days of empire had faded, but the European powers had retained colonies in Africa, and before the war France had occupied a large portion of northern Africa, including Morocco, Algeria, and Tunisia. These colonies (technically, Morocco and Tunisia were protectorates, but it amounted to the same thing as far as we were concerned) sat now under the control of Germany's puppet, Vichy France. The troops there were all Vichy. While we expected at least one high-ranking commander would defect to our side and many troops might not fight, there was no way to know how stiff resistance might be.

Beyond those countries sat plenty of German and Italian troops, and they certainly would fight. Libya, just to the east of Tunisia, was an Italian colony. Bordering Libya to the east was Egypt, which though in theory independent had been occupied by Great Britain since 1882.

The Axis and the British had been fighting in the Libyan and Egyptian deserts since June 1940. The British, under General Harold Alexander and Lieutenant General Bernard Montgomery, had recently withstood a push by the Germans eastward; they were now in the early stages of a drive against the Axis aimed ultimately at Tunisia. The idea of our operation, code named Torch, was to take over the weakly defended French colonies and come eastward toward Tunisia, pressuring the Germans as the main British offensive continued to the east.

Besides defeating the German forces and neutralizing the French Vichy assets, Torch was intended to help tie up German troops and resources, hopefully draining them from the fight with the Soviet Union in Russia. That was a strategic play far beyond the concerns we had on the ships as we sailed. Always for us the war was an immediate affair; the only strategy that counted was the one that kept you and your buddies alive.

Nearly 60,000 men and some 250 ships were involved in the overall operation. They were divided unevenly among three task forces, west to east, with landing areas near Casablanca, Oran, and Algiers. The 16th Infantry was part of Center Task Force, a group that included the 1st Ranger Battalion as well as our sister regiment the 18th Infantry. Starting from beaches along a fifty-mile stretch of the Mediterranean, the main objectives included two airfields as well as the city of Oran in French Algeria.

With a population of about two hundred thousand, the city was defended by nearly seventeen thousand Vichy French and colonial troops; besides coastal defenses, there were destroyers in its harbor and an estimated one hundred aircraft at the nearby airfields. There was hope that the city and nearby troops would surrender—the Allies had been negotiating with officials there, who were none too keen about their German overlords—but with no agreement as we approached, we couldn't count on hope.

The 16th Infantry Regiment and its roughly 3,300 men formed the core of Regimental Combat Team 16, designated to land near Azrou on a peninsula some twenty miles by air east of Oran. Our battalions were on two ships, the *Bedford* and the *Warwick Castle*; my brother and his unit happened to be on the latter. Our escorts included one battleship, three aircraft carriers, three cruisers, and a number of destroyers, all of which would provide us with firepower when we landed. The first wave was set to hit the beach—"H-Hour"—at 0100—1 A.M.—on November 8, or as we wrote in the military, 8 November 1942.

IT STARTS

Some units, including those from the 16th, did get off on time. Others were hampered by the usual confusion and disorder that arrive to baffle nearly every military operation a few minutes before they step off. The waves were stronger than anticipated; there were sandbars where none were expected; the men steering the landing craft got confused about where they were supposed to go.

We went ahead anyway.

The landings took place on both sides of Oran; the idea was to form a noose around the city, eliminating forces on all sides before tightening the rope and entering. My battalion was tasked to land near Azrou, on the eastern side of Oran. We were held in reserve, reinforcements who would be used after the beaches were cleared, or in the event plans had to be changed.

The initial assault was under almost ideal weather conditions, despite some wind and temperatures ranging from 40 to 50 degrees. The surf was high, with six rows of breakers, some five feet high; wading through that, especially from a couple of hundred yards out after your boat hit a sandbar, wasn't a lot of fun. But the first wave was lightly opposed; by the time I went ashore around two o'clock in the afternoon, what little resistance there might have been on the beach was over.

It was a different story inland to the south, where we encountered a hell of a fight from Vichy French and colonial forces still loyal to the Germans. They continued to resist, fighting at times in isolated pockets and even counterattacking. But our men pressed toward Oran, reaching the outskirts early in the morning of November 10; by one in the afternoon our troops were in the city.

Moving our battalion aid station, manned by myself and four other medics, to keep up with the fighting took some imagination as well as legwork. Poor planning and difficulty in getting the landing craft to the right spots played havoc with our supplies and logistics. We were supposed to have vehicles to move our medical gear, but the Jeeps were nowhere to be found. So I improvised.

The area east of the city was a farming community, and there was a small farm near the road. I found the owner, who seemed to understand English—or enough to say that we could use one of his horses and a two-wheeled cart to move our medical equipment. I gave him a receipt, and we used the cart to get our medical chest and supplies all the way into Oran, which by that time had been occupied by our troops.

While the resistance had been light, we did have casualties to care for. I remember putting only two fellows into the cart, neither with life-threatening injuries. Others in the 16th weren't as lucky. Private First Class Albert F. Smith of E Company may have been our first man killed in action (KIA); his remains lie today at the North Africa American Cemetery and Memorial in Carthage.

The regiment's casualties totaled twenty-five KIA, with seventy-nine wounded. That is the terrible math of war—even in "light action" or "small resistance," as the fighting at Oran is often described in the history books, real people die.

For a first engagement, and especially a first landing under fire, we had done well. We'd been prepared for more casualties; it was a blessing that we didn't get them.

The first casualties my guys treated were, fortunately, not blown apart or very seriously wounded. So it was a relatively easy introduction to the realities of combat.

As for myself, my earlier training at the hospital had taught me how to distance myself from my patients so I could work. Now I saw what death during war looked like, at least in some of its more common varieties. I would see much more as we went on.

The one thing that I was surprised and concerned about soon after we landed was the power lines. Lines had fallen into ditches along the road; electricity still ran in some of them, and we found one or two men who'd unknowingly stepped in the ditches or jumped in because of gunfire. They'd been electrocuted, something we had neither encountered nor thought about. The problem was that you couldn't tell from looking at a ditch or a wire whether it was a hazard or not. And the only indications when you'd see someone in the ditch were scorches on the body and clothes.

Our Jeeps came in a few days later, and we returned the farmer's cart and horse. I think he was the most surprised man in all of Africa when we turned up with it.

The landings in Africa are often overlooked in histories about the war, but they were important for many reasons. We learned a lot from them. There were very specific lessons, like the fact that syrettes containing premeasured doses of morphine were the best way to administer the drug in combat, and that transfusions with blood plasma could save a large number of lives on the battlefield. While the nature of plasma had been known for centuries, it was only first used during our war.

Most important, we found out what it meant to be shot at for real.

ORAN

I don't remember seeing any civilians on the road into the city. I guess when they caught wind of what was coming, they sought cover or fled. Once the fighting was over, though, they returned or showed themselves. Life in the city—well, I can't say it returned exactly to normal, but people began going about their business again. This would become a routine not only throughout Algeria, but also Africa and beyond.

Our battalion was assigned to provide security in the city. We set up shop in an old French fort. Our role as medics meant we were responsible for the general health and well-being of the troops. That turned us into inspectors—of food and sanitation, and what back in the States would be called "houses of ill repute."

Whorehouses would be the cruder but more accurate term.

I went out on these inspections, escorted by the local gendarmes. Most of these places, which had apparently been functioning for quite a while before our arrival, were located on a single street, with a handful of others sprinkled nearby. We'd inspect the houses themselves for cleanliness, then line up the girls and check them for sores and other telltale signs of disease. They all had, or were supposed to have, doctor's papers declaring that they were healthy. Without those, they couldn't work.

I remember designating one house off-limits, but otherwise every place we checked was in order.

I also had to designate one home for officers, where enlisted men would be barred. That was easy. The officers got the house with the ugliest women.

During one of the inspections I came across a woman who was an American, and easily recognized as such from her clothes as well as her accent. We chatted a bit; she said she'd been trapped by the war, and turned to prostitution to make money. And apparently she did very well. She told me that she was saving her earnings, and when the war was over would return home—Chicago, as I recall—and no one would know.

It wasn't my place to argue.

It may seem shocking to some, but prostitution was an accepted and expected part of the war, and even before. In fact, I remember one town near Fort Riley in Kansas where there were so many prostitutes, you could hardly drive through without being propositioned if you were in uniform.

While we didn't inspect such places where there were functioning civil authorities like in the States or England, we did set up stations to check men for venereal diseases. "Short arm inspections"—you can work out what that means—were a necessary part of the job. We set up what were called "pro stations" where men would come in for inspections, which could include not very pleasant probes.

Kits with sulfa drugs as well as condoms were routinely issued to GIs throughout the war, along with pamphlets that probably told them everything they already knew about the dangers of diseases such as gonorrhea and syphilis and how to avoid them.

I did not partake of any such services. There was never a case overseas that I went to bed with any woman.

For the record.

With the city secure, the "funny boys" came in—headquarters people, quartermasters, and the troops assigned to work with the local government. We called them funny boys because unlike us, their uniforms were always clean.

Once they were in place, the 16th began moving east with the rest of 1st Division. We took a several-hundred-mile "jump" to Guelma, Algeria, some forty miles south of the coast.

As big as that movement was, the Big Red One and its units

were still well west of the Eastern Task Force, which was driving from its landing area in Algiers toward Tunisia. Under the command of British First Army General Kenneth Anderson, this force of British and Americans aimed to take Tunis, where it would hook up with the British troops under Montgomery. From there, the Allies would push the remaining Germans northward toward Bizerte, where the last survivors would be captured or destroyed.

Life in the villages we passed through seemed primitive, even by Alabama standards. The farmhouses were small huts, and if there was a pen for animals, it generally opened into the living structure, something that struck me as odd.

This was also my first exposure to Muslims. One thing we quickly realized—women here had few rights and seemed to be treated with little respect. They walked while their husbands rode, and even when walking together, they stayed several paces behind, as if in the second rank. Women had to wait for the men to finish eating before they could start; they got the leftovers. A man could have as many wives as he wanted, another foreign concept.

We were careful not to insult the locals, or even get into too much conversation. I'd been told that if I admired something a local possessed, he might very well give it to me—and expect the same from me. I didn't have much to give; all I could think of was having to turn over my wedding ring. So I kept my distance.

While the local French government had mostly stopped resisting and was now officially supporting the Allies, the Germans were desperately moving men into position to oppose us. Their forces were relatively small, but the Germans had good air support, and in some instances still had French and colonial soldiers willing to fight for them.

When most people think of northern Africa, they picture flat, dry land. Much of it is that way—but the area that the Eastern Task Force was moving through into Tunisia was neither flat nor particularly dry. The hills and mountains were inundated with rainstorms, turning the roads and fields into mud and hampering our advance. The Task Force had its own problems with logistics, and despite forcing the Germans to retreat deep into Tunisia, bogged down in December short of its objectives.

As the spearhead stalled, so did the entire offensive. That stopped us as well. While we didn't have nearly as much fighting in our sector, we still had our share—which was more than enough for some guys.

SELF-INFLICTED WOUNDS, COMPANY MEN

Sometime after leaving Oran, our staff sergeant and one of the other medics went out to check on some of the companies. While there were no large enemy units in the area that we knew of, there were still places where machine guns or snipers would set up to oppose us. We were constantly having to check on small huts and villages to make sure the enemy wasn't lurking there.

A few hours later, the Jeep roared back into the aid station where I was.

"Staff sergeant's been shot!" yelled the medic at the wheel.

We got him out and brought him over to the treatment area. He'd been shot in the foot, and it was obvious from the powder burns and condition of the wound that he had been shot at very close range, almost surely by himself.

There's no record of his casualty in the unit medical listings that I still have, so I'm not sure how the incident was officially recorded, if it ever was. The result, though, was this:

He went home. I was promoted to staff sergeant.

The truth was, the change in the insignia on my sleeve wasn't a big deal. I'd already been doing a lot of his job.

Our structure in Africa was essentially the same as we'd had before combat. The one big change: where once we'd had only a single doctor with us, now we often had two. The extra doctor would rotate in from the regimental medical team, in effect the echelon above us. Dr. Morchan remained with us throughout most of the campaign, and would continue to do so through Sicily. I trusted him a great deal, but all our doctors knew their jobs and did their best.

And they pretty much left me to do mine.

I knew most of the rest of the medics very well, since the majority had been with the unit since 1940. I did things a little differently than some of the other sergeants. In most medical units, the men had very specific roles and stuck to them throughout the campaign. A company medic was *always* a company medic, a litter bearer always a litter bearer. I rotated my guys as much as possible. Besides keeping them fresh, I was spreading the risk.

There are no safe jobs in the war zone, and the front line is a place where life expectancy can often be measured in minutes, not decades. But one of the most dangerous jobs in the army is being a company aid man—the medic who rushes to help a freshly wounded soldier in battle.

The reason is simple: you can't dig in. You have to be seen. You have to keep yourself visible so the infantry guys know where you are. They have to know that if they go down, someone is going to rush to their aid.

I'm not saying men won't fight without that knowledge. I am saying that they will fight harder and better if they know "Doc" will rush up to help them if they go down.

Picking a company aid man was more art than science. He had to be qualified to give first aid, obviously. More than that, he had to have the stamina to keep up with the rest of the rifle company whether they were marching or on the battlefield. He had to be able to mix in with the infantrymen he worked with; be liked. Getting along with different types of people under potentially catastrophic conditions is not easy. He had to be brave under fire, and even rarer, able to concentrate on his job despite the worst chaos around him.

I remembered to say he had to know first aid, didn't I? That was his primary job. He'd assess the wounds, decide how bad off the injured man was. If the soldier was bleeding, he tried to stop it, bandaging or tourniqueting as necessary. He'd give the soldier a shot of morphine. Then, if the unit was in heavy action, the company aid man would move on to the next casualty.

Maybe he'd get a chance to duck in between. Often not.

Litter bearers didn't have it easy, either. They, too, were often exposed to gunfire. Working in pairs, they would take the wounded man from the front line and bring him back to the aid station, where the doctor and a small group of medics would work on him.

While ideally it was away from direct gunfire, our aid station had to be relatively close to the fighting. That meant fifty yards on occasion. Two or three hundred was probably more common, but the terrain always had to be taken into account.

The aid stations weren't necessarily safe. Most often, they were subject to artillery and mortar fire; occasionally they were within range of a rifle bullet, generally one that missed its mark. But they were comparatively safer than the other positions. So I rotated the guys and gave them a bit of a break. Medics who worked in the aid station with me would go out as company guys while the others took their place for a while. I felt it was fair that way, so no one was constantly exposed to direct fire or artillery, hand-to-hand fighting or whatever.

We all have our own level of stress; every job in the military does as well. Medics are unique. I won't compare our stress to that of an infantry rifleman; all I can say is both jobs are very intense experiences under combat.

They became more so as we moved east toward Tunisia and the fighting in Africa progressed, because the German sharpshooters started aiming at medics.

The red crosses we wore on our helmets and armbands now made us prime targets. The enemy was aiming to cripple our troops' morale by taking us out. It was a deliberate policy, a war crime. Snipers gunned down a number of us before Command issued a warning and most of us removed the armbands.

We were also authorized to carry weapons, something we hadn't done to that point.

I put my Red Cross symbols away and holstered a .45, just in case.

LEARNING TO KILL

The 1st Division was sent into action just before Christmas in a fresh effort to break the German line. This offensive bogged down within a day, tormented by rainstorms as much as by the German defenses. It was the latest in a series of setbacks, especially disappointing given the earlier momentum following the landings.

We didn't know it in the field, but back in Washington the Allies' slow progress, and especially the sluggish showing by the Americans, set all kinds of alarm bells ringing. This eventually led to big changes at the top of our chain of command. In the meantime, the division took up positions in the Ousseltia Valley, part of II Corps under Major General Lloyd Fredendall.

Our II Corps—pronounced "two core"—was south of Anderson's force, which remained the key fighting group on the west side of the Germans. Our area of operations was a mishmash of hills, ridges, depressions, and valleys in and between the Atlas mountains and surrounding hills. While the passes east were very limited and easily defended, if II Corps were to cut through them and out of the mountains east, we would have been able to split Tunisia roughly in half, cutting off German supply lines and making it difficult for them to escape if Montgomery's attacks became too intense.

The Germans recognized that and decided to do something about it. We got beat pretty bad in a series of battles in the Tunisian hills and passes that January and February. The climax of those battles took place at Kasserine Pass. Our forces ended up turning back the German offensive there, but at great cost and not before the German attack had accomplished everything its architect and commander, Erwin Rommel, wanted. Not only did the Germans end up with good defensive positions, but they had taken the fight out of our commander, and not a few of the men.

Rommel is a legendary general; he needs no kind words from me to make his reputation, and he'll get none. Plenty of people in occupied France later felt his cruelty; there are monuments to the civilians his troops killed there. But there is no denying he marshaled his forces cleverly, and whipped us good in Tunisia. When he was done with us, he turned his attention back to Montgomery, whose British force was coming at him from the south, still aiming to push his troops into the sea.

One good thing came out of these losses: General Dwight Eisenhower appointed George Patton to take over command from General Fredendall in early March. General Patton picked Omar Bradley, who'd been sent by Marshall to find out what the devil was going on, as his second in command. They were an improbable team, as different as any two generals in the war, but they ended up doing a fine job.

Why did we do so poorly from roughly December to late February? By that point we had mostly solved our logistics problems; supplies were flowing regularly. You never have enough in a war, but things were hardly as bad as they had been when we first landed. Our forces outnumbered Rommel's both overall and in most engagements, though he and his underlings took whatever momentary advantages they could.

One of the biggest factors, in my opinion, was our inexperience. Not only did we not really know war yet, we didn't know how to kill.

It's more than shooting someone. It's not something you learn in your head, not a math equation or an instruction about how to wire up a switch. It's knowledge you need to get into your bones, into your heart. It's a harsh thing, but without it, you and your friends are dead, your battle is lost, and what you came to fight for is forfeit.

The Germans had been fighting for years. They were warriors. They had good equipment, sometimes better than what we had, though generally not as much. Most of all, they knew how to kill. They weren't reluctant to do it.

You're taught all your life not to. *Thou shalt not kill.* Unlearning that is hard.

At some point, you get it in your mind that you are either going to kill or be killed. You see your buddies being slaughtered and realize the enemy is trying to kill you, too, and you've got to do the same thing to them if you want to survive.

But for many men, there's still hesitation. Until you can overcome that, both with experience and good leadership, your army will always be inferior.

It was the hardest thing to learn. And we had to do it real quick, or the war would be lost.

We had excellent leaders at the top of the 1st Division, starting with our commander, Major General Terry Allen. General Allen was one of the best-loved generals in World War II, and he's the one I think did more than anyone to help the guys learn how to get in there and start fighting the way they should. He'd been in World War I, where he'd been shot, won the Silver Star, and learned how to lead. He was a true soldier's general, a man who sat on the ground next to you, who had an aggressive combat style.

A man you could imagine calling Terry, rather than General Allen—though you never would.

His second in command was Brigadier General Theodore Roosevelt, the son of President Teddy Roosevelt. He was another action guy who'd fought as an officer in World War I, where he'd been gassed and won the Distinguished Service Cross (DSC), the army's highest award, one step below the Medal of Honor, for his bravery. Rejoining the army just before the outbreak of World War II, he was given command of the 26th Infantry Regiment, his old unit in the 1st Division, then promoted to be Allen's assistant division commander. Like Allen, he liked to lead from the front.

Both generals realized our shortcomings and worked to overcome them. Once the fighting got tough, they encouraged and sometimes goaded the combat officers to do more. I ended up meeting a lot of generals—my aid station was often the place they'd stop when touring the front. No one I met—Bradley, Patton, Eisenhower, you name him—was the measure of Allen or Roosevelt.

CALL ME DADDY

I found out I was a father at the end of January 1943 while I was in the Ousseltia Valley of Tunisia, west of the German lines. While Estelle had written to me about her pregnancy when I was in Great Britain, the news of our son's birth still caught me by surprise. He'd been born January 16; almost two weeks had passed before the news arrived in the form of a Red Cross volunteer who came to our unit and told me.

My response was not carefully considered.

"It's about time you got around to telling me," I snapped.

Not very understanding, I know.

Estelle had named the baby Arthur, after me. To be honest, I wouldn't have made him a "junior." I know it was intended as an honor, but it's the kind of thing that causes a lot of confusion later on in life with bank accounts, official records, and that sort of thing. But I had no say in the matter, being out of touch.

I was able to write a letter home, telling Estelle that I'd heard we were parents. A few months after that, I got a letter from her with the details of how big our son was at birth—about six and a half pounds. I expect he'd gained two or three times that by the time I got the letter.

I wondered how big he'd be when I got home and finally met him.

By mid-March, command of the Allied armies had been rejiggered and the offensive had started up again. One morning as our companies were moving up, I took a Jeep out with one of my medics, Bud Hays, to scout for a place we could locate the aid station. There were farms amid the hills, and water flowed through the wadis, or seasonal streams, though the day itself was dry. Sitting in the passenger seat as we drove across a field, I concentrated on the terrain, looking for a spot that would be close to the guys at the front but sheltered, if possible, from German artillery, ideally a little nook at the side of a hill.

BAM!

The world turned upside down.

The Jeep flipped and I went flying. Our left front wheel had hit one of many land mines planted by the Germans. I was lucky to be thrown free. I landed on my knees, banging the hell out of them and drawing blood, but not breaking any bones. Hays wasn't so lucky, though he, too, got off a lot lighter than he might have, breaking his arm rather than his neck. The explosives themselves hadn't hurt either of us.

The Jeep, though, was ruined. We gathered ourselves and walked back to our post. My wounds were the very definition of "walking wounded," I suppose, not bad enough to report, let alone keep me from working. We'd all get banged up from time to time; we treated ourselves and moved on. Only the very serious injuries were recorded.

It was different with the soldiers we cared for. You were always concerned that an infection might develop or the injury might prove more serious than first suspected. So we kept records that would allow treatments to be followed, and to make sure nothing came back to bite us.

Our tangle with the land mine happened right around the time of the Battle of El Guettar, when the Big Red One went up against the Germans' 10th Panzer Division. The Germans nearly overran us in the El Guettar Valley in Tunisia; their tanks rumbled close to General Allen's headquarters at one point. But artillery and tank destroyers cut down the enemy spearhead, and for the first time in Africa, American troops defeated a front-line German unit. The Nazis fell back; while we were too battered to take quick advantage, we had finally turned the momentum around.

On maps, battles are large and small arrows, dotted lines, terse descriptions. On the ground they're flipped-over Jeeps and busted tanks. The thick arrow might represent several weeks of fighting, during which a unit might be cycled on and off the front line several times.

The maps can't show things like the mud that crusts on your boots or the dirt that coats your skin. It certainly doesn't show the blood that cakes on your trousers after you've cleaned a dozen wounds.

This was the middle of a three-month stretch when I went without a shower. The best I could do when we had a lull in the fighting was take some water out of a canteen and pour it into my helmet, then use my undershirt as a washcloth. Lord knows how I smelled when I was done, but it probably wasn't any worse than the other fellows.

THE PIPE

We were slowly getting control of the skies, thanks to the airfields we'd captured back west. But the Germans were still very active, attacking with a variety of planes. They were also sending reconnaissance aircraft over regularly to see where we were, helping them aim their artillery and ground attacks.

One afternoon in the valley, we heard the angry rasp of an aircraft sprinting over our aid station and the surrounding olive grove. I looked up in time to see a twin-engine P-38 Lightning swoop into a dive.

The army air corps P-38s were awesome fighters, nimble and fast, and rightly feared by the Germans, who called them fork-tailed devils because of their unique boom tails. In capable hands, the American fighters or "pursuit planes" were more than a match for anything the Germans had in Africa. In this case, the Lightning had an extreme advantage—its prey was a light German observation plane we called a "Stork."

If you've ever seen a Piper Cub or even a high-winged Cessna, you can visualize the basic silhouette of a Stork, whose proper name was the Fieseler Fi 156 Storch, German for stork. Unarmed, it was a narrow plane with wings that seemed disproportionately long. The pilot and whatever passenger he might have (often none) sat in tandem, one behind the other. As an observation plane, it was designed to fly low and slow. These flight characteristics were an asset near the front lines, since they allowed the Stork to take off and land from roads or even fields with no need for a long or well-prepared runway.

But it also meant that it had a low top speed and was no match for the faster, well-armed Lightning.

The Stork's pilot knew he was flying for his life. He dove low, wheels nearly scraping the top of the olive trees. The Lightning followed, but couldn't quite get low enough to line up and fire. The Stork banked hard and the Lightning flew by, cutting speed and twisting back behind its target.

Again and again the two planes pirouetted above us. For quite a while, the German managed to turn or dive or jerk upward a split second before the P-38 pilot could fire.

Then his luck or skill ran short. A quick burst hit the German and the Stork crashed a hundred yards away, maybe closer. The Lightning roared off, and the air was suddenly quiet again.

I ran for the plane. Now that it was down, the pilot was no longer an enemy—he was just a man who needed help.

Though shot down and mangled in the landing, the plane was intact enough that I could get to the door at the side and pull it open. Blood was everywhere. The pilot had been hit by the Lightning's bullets or some shrapnel, and probably broke some bones in the landing, but he was still alive. He looked at me, dazed but awake. Worried that the plane would explode any second, I reached in to grab him and haul him out.

He started talking, extending his hand toward me.

There was something in his hand. I jerked back, then realized it was too small for a gun.

It was a small tobacco pipe. He seemed to be pleading with me to take it.

Was it a reward for helping him, a bequest for someone?

He kept talking in German. I finally took the pipe. Before I could make sense of anything, before I could ask, even in English, what he wanted me to do with it, he slumped down, dead.

I stuck the pipe in my pocket and walked away. There was nothing more I could do for the man in the plane. He was back to being an enemy.

Later on, I tossed the pipe in my med kit, where it stayed for the duration of the war. I guess you'd call it a memento now; it's one of the very few I have from the war.

A few years after the war, I briefly took up pipe smoking. I used the German's pipe a few times, to see what it was like.

Every so often, I take it out of the case where I keep it. It's a fine piece, nicely carved with a horse and hunting scene on the bowl. Was it a family heirloom, passed down for several generations? Did he want me to give it to someone? Or was he offering it to me as a reward for helping him? Each time I hold it in my hand, I wonder what he was trying to say.

When I think of him now, he's just a man again. Time does change some things for the better.

DEATH IN THE FAMILY

Now as we fought, the Germans were always close by. Even at points when the two sides weren't charging at each other in the sector, we regularly fired at each other, either with artillery and mortars or, in places where the soldiers were close enough, direct gunfire. Every day, without end.

The Germans had a large number of 88-millimeter guns, originally designed as antiaircraft weapons. First put to use against tanks when the Germans supplied General Francisco Franco's nationalist forces in the Spanish Civil War, they had come into their own with Rommel. Famous as the "88," the cannons were hated and feared at the same time. They were also so famous that to us, pretty much every time we were shelled with something other than a mortar round, we'd say it was from an 88. We didn't stop to measure the shell. There were so many 88s that the odds favored them anyway.

Because the aid stations were in fixed locations close to the front lines, they were constantly in danger. Any Geneva Conventions protecting them or the wounded had no force on the battlefront, and even if the enemy wasn't deliberately aiming at us, the slightest error when firing could easily put us rather than another unit in the bull's-eye.

Toward the end of March, one of my medics by the name of Blackwell got hit with shrapnel or shell fragments during an attack. The wounds weren't life-threatening, but they were bad enough that we had to send him back to get patched up. He returned to us on March 30, full of energy and glad to be back. I suspect he'd talked the doctors into letting him come back early; like the rest of us, he was always trying to do as much as he could, going beyond what would be expected.

That day, March 30, 1943, we were in a lull, with minimal fighting in our sector. A Jeep came up to the aid station with thermos cans of food, and men from a nearby company came down for chow. We were very close to the front line but shaded from it by the hillside, which gave us a little protection.

Or so we thought.

Something whistled overhead as the guys bunched up to get food. A German shell hit with impeccable timing; Rommel himself couldn't have aimed or timed it better.

Blackwell was among the men in the little clump that got hit. His wounds were fatal. He was our first KIA, the first death in our family.

I blamed myself. I was in charge. I felt I could have done something to prevent his death.

What could I have done? How was it my fault?

Those are logical questions, with logical answers. Nothing, and it wasn't.

But they completely miss the point.

What I felt was guilt, and responsibility. Emotion. Not fact, and yet as true and real as any scientific equation.

I know now, logically, that it wasn't my fault, that it was bad luck, and a German shell after all. But that feeling then was real, and I can still taste it in my mouth talking about it, all these years and deaths later.

I'd been wounded in the attack as well, but not badly enough to keep me out of action, nor was the shrapnel that hit my left shoulder a few days later. One set of those wounds—probably the first—was recorded as a bare note lacking any details except that I had been treated for a battle injury; according to the records, it happened on April 1. Whichever it was, it earned me my first Purple Heart.

The Purple Heart is not an honor you really want when you're signing up for the army. After you leave the service, it signals that you were in combat, and so demonstrates that you put yourself in harm's way to do your duty. But getting shot or hit by artillery is not the object of war. Doing that to the enemy is.

While I was "earning" my Purple Heart episode, our troops were clearing a key hill held by the Italians near El Guettar, a strategic point protecting the Axis flank. Tactically, the 1st Division and its sister troops were now threatening to cut behind the Germans, who were already being battered by Montgomery to the south.

In the immediate moment on the ground, though, this wasn't clear to us. The forces scrambled in the hills, fighting back and forth as if they were involved in a chess match where the slightest advantage in position might make all the difference.

Except this chess match was being played with guns and explosives.

ON THE HILL

My company men had been pushed hard, living with their units without much of a break since we'd moved into the valley. Things got even hotter for them as the Germans realized they were close to being trapped. As Rommel's force retreated, they fought desperately against us. They'd been ordered to hold out to the last, and most of them took that order very seriously.

One night I decided to leave our aid station to check on the company that was fighting on the hill above, not more than two or three hundred yards away, if that. They'd had heavy action during the day and I wanted to see how my medic was doing.

It was just getting dark. I could see well enough to pick my way across the slope, heading for the peak where the company had dug in. I stopped for a moment, getting my bearings. I saw one or two of our soldiers about two hundred feet away, on the other side of the hill.

Just then I heard a scraping noise above me. I stepped back against a boulder, trapped momentarily as a figure loomed above. He had a rifle in his hands, bayonet fixed at the end.

He yelled something, and I realized he was German.

Then he leapt down, charging at me, blade first.

FIVE

My Life in My Hands

INSTINCTS

I ducked to the side like a matador, grabbing for the rifle barrel. I caught hold of the wood under it, but the German pulled back quickly, and the blade of the bayonet sliced through my hand.

I tucked to the side, still trapped. Time seemed to move in two different directions simultaneously, quick and slow.

The German stumbled, gathered his strength, then charged toward me.

My fingers were sliced, my field jacket cut. I was still alive, though. I dodged.

A thousand thoughts and none flashed inside my mind.

Without really thinking, I reached to the holster I'd been carrying for weeks and pulled out the Colt. As the German charged again, I fired point blank into him, twice.

I ducked as he fell off to the side, tumbling a few feet down the hill.

How did I pull the trigger with my finger half sawed off? How did I have the presence of mind even to draw the weapon, let alone aim it?

Questions I still ask.

I bandaged my hand, stopping the bleeding, then looked at the German.

Dead. He was the first man I'd killed in combat.

Clearly an enemy.

But he was about my age. Maybe not that much different in many ways.

I was no more than twenty-five yards from the GIs on top of the hill, but they hadn't heard the fight. After making sure my enemy was dead, I walked back down to the aid station. By then it was dark. The air tent was lit by an oil lamp, with the light just bright enough to show how red the blood on my hand and gun was.

"What the hell happened to you?" the doctor on duty asked.

"I got in a fight with a German."

I gave him the details. Our litter bearers went out to get him. When they brought him back, we checked for his ID. German dog tags were oval-shaped, and in two parts, so that one could be kept with the body and the other sent for records. As I looked for his wallet, I found a photo in his pocket. It was of him and two women.

I guessed they were his wife or girlfriend and a sister. The sort of photo I might have.

There was writing on the back, but nothing that explained who or what the people were. Without thinking, I tucked the picture away. Maybe at that moment I intended to track down the family, or more likely, thought I could give it to whoever would be responsible for repatriating his body. But none of that ever happened, and today it's another of the very few mementos I have.

Along with the knife he'd tried to kill me with.

I don't think I was scared during the fight. There was so much that happened in those few moments, with the importance of every inch of movement multiplied. But as soon as it was over, I broke out into a sweat.

A SILVER STAR

My hand was chopped so bad I had to go back to the regimental aid station to get thoroughly stitched up. My injuries had to be recorded, which brought me another Purple Heart. Then they told me to go back and get the other hand shot off.

Not exactly. But I went back happily anyway.

Over the years, the exact details of the fight have softened, but I can still look at my hand and wonder how I was able to get out alive. And I can wonder especially about the man who tried to kill me.

The German was either some sort of sniper or a straggler. How long he'd been there was impossible to tell. Very possibly he'd survived an earlier battle, maybe been wounded. Undoubtedly he was out of ammo—a good thing for me. Why he chose that moment to come out of hiding I'll never know.

Possibly, he was as surprised to see me as I was to see him. Maybe he thought he had a clear run down the hill back to his lines. Or maybe he'd been waiting for hours to find one American to take a last revenge on before he died.

The front lines at that time twisted all through the hills. You couldn't necessarily tell who held which piece of ground if there was no fighting or groups of soldiers nearby. Not only did the small units at the forefront of the fight move around, but the geography increased the uncertainty. Physical barriers or markers like a river were rare. Many times a road would belong to whoever was near it. If no one was firing at you, you might assume you were behind your own lines, only to have a very rude awakening.

Three days after the bayonet attack, the 1st Division and the rest of what was now called II Corps made a heavy attack against the Germans. Code named Operation Vulcan, this drive was like a spike in the side of the German forces. It was a slow, bloody push east, hammering the Germans as they retreated from the British in the south toward Tunis.

The 1st Division moved up the Tine River Valley in north-central Tunisia, making progress in fits and starts.

We'd become a different army by then. Months of war and getting our noses bashed in had taught us a lot—most important, how to kill.

Besides our division, II Corps included 1st Armored and the 9th and 34th Infantry Divisions. Patton had secretly returned west to prepare for the next operation, but Bradley was just as aggressive.

Still, the Germans fought back hard. They would constantly counterattack, or launch what were called spoiling attacks aimed to throw off our plans when a fresh push was in the wind. A spoiling attack was like a light jab; the Germans hoped it would cause us to rush to that spot, postponing or even calling off the main assault.

It was a nice trick. We didn't fall for it too often.

I was back with my unit as the offensive got under way. By

April 28, we'd reached the outskirts of Mateur, which was west of both Bizerte and Tunis and about equidistant from each, at least as the crow flies.

The Germans began pounding our units with artillery shells. The aid station itself was close to the shelling, so we pushed it to a safer spot. Then I went up in the Jeep to help one of the nearby companies gather the wounded.

Between the shells falling all around and the twisting terrain, I didn't realize the road took me right through the path of a German advance. Ignorance was bliss—I got to the company and piled five guys in and onto the Jeep before taking off back to the aid station. I made it without a problem, or so I thought. Some bearers took the wounded and I was off again.

Going back up, I realized I was being fired at. Bullets flew past the Jeep. The artillery barrage had been meant to soften up our defenses for an attack, and here it came—right across the road I was driving, directly into the heart of where my guys were.

I found five soldiers lying on the battlefield as the Germans charged past. I got the GIs into the Jeep, then hauled out of there, racing down the roads as quickly as I could.

There were a lot more wounded back at the position; men were lying on the ground where they'd been hit by shells or shrapnel or gunfire. The Germans were already through the position, charging toward the survivors.

The way the roads were laid out in the hills, the only way to reach the men who'd fallen was to go back the way I came.

So I did.

This time bullets hit the Jeep. Somehow I made it to the field. I could only find two more men on the ground before I realized it was time to make a U-turn.

"This is the last time I can go," I told myself.

By now it was all too obvious I was behind the enemy advance. At least one German machine gunner began throwing lead at me as I drove away. The only thing that saved me was the sheer confusion of the battle. Everything was exploding—artillery, mortars, rifle fire.

I made it to the aid station. There was no time to count the holes in the Jeep—I helped get the men from the vehicle and went to work on their wounds.

I guess they were impressed when they heard the story back at Division Headquarters. Three months later, I got a citation signed by Terry Allen himself. Sandwiched between the words "Confidential" was a message that read in part:

Under the provisions of AR 600-45, as amended, Staff Sergeant Arnold R. Lambert, 7006617, Medical Detachment, 16th Infantry, is awarded the Silver Star for gallantry in action.

BOUNCING BETTIES

You would think that after so many battles, and so many bullets flying close, death would be something on your mind.

It wasn't on mine. I didn't think about dying, or even that I could die.

I don't mean I wasn't scared at times, but that was more an immediate, emotional reaction to things. Sitting around thinking about my death, how I would get shot or blown up, what my parents would think, who'd come to my funeral—none of that was on my mind.

It would have been realistic or at least understandable to have those thoughts. And truly, I knew people were going to die. But death was abstract. Or maybe more accurately—it was going to happen to someone else, not me.

Never me.

I have strong memories of different things, bits of battles unmoored from what came before and after. That loss of context must be what happens when you grow old, the tax you pay for living a long life.

I'm thinking it was around this time or not too long before that I went with one of my men to scout ahead for an aid station. We were doing that a lot now; it meant the battalion and the division were making real gains and moving up.

There was a tank battle going on somewhere ahead, and we were in a big flat area. We saw a guy walking in a field off the side of where we were. Suddenly, there was a puff of smoke. He'd stepped on a mine, a Bouncing Betty, I think.

The Bouncing Betties were nasty pieces of work. They were S-mines to the Germans, *S* for *Schrapnellmine* or *Splittermine*. You'd step on them, and the mine would spring up into the air, igniting right about groin level. It was designed to spray fragments of the metal surrounding the explosive into your crotch, castrating you but keeping you alive.

At least that's what we thought. You can imagine what it would do psychologically, and not just to the man who was maimed.

We saw the dust as it settled and stopped. My driver started to get out to retrieve him.

"I'll go," I insisted. We argued very briefly, but I made it clear

I was going—I wouldn't send anyone else in my place, no matter the danger.

I got out and found the GI's tracks through the dirt. I followed his tracks to where he lay. True to its design, the mine's shrapnel had caught him in the groin area and around the legs. But he was alive, and I could tell he would live if we treated him. He put one arm around my neck and I picked him up, carrying him like a baby back through the tracks that took me to him. We put him in the Jeep and hustled back to the aid station.

I never figured out why he was alone there. Maybe he was catching up to his unit. Strange things happen in war, and you never know how or why.

High-ranking officers and VIPs often stopped at our aid station, even when things were hot, since it was about as close as you could get to the front line without—usually—exposing yourself to direct fire. General Allen would come in a lot.

But the highest-ranking general who came in wasn't there to observe; he came to get patched up.

It was Lieutenant General Lesley McNair.

By this point, McNair had had a long and distinguished career. He'd been in World War I and had made brigadier general at age thirty-five—then and now incredibly young for such an important post. Just before the war, he'd been made chief of staff at Army General Headquarters, or GHQ, and now was head of Army Ground Forces, posts that made him responsible for increasing the size and training of our forces, and was largely responsible for organizing the army's structure and the way we fought.

McNair had come to Africa to get a firsthand look at what was going on. While here, a German shell had hit nearby, killing a man as well as slightly wounding McNair.

It excited us a bit to treat someone of that stature. His wounds, fortunately, weren't very severe. A little sulfa, a bandage, and he was fine.

I didn't work on him, but I did talk to him. In times like that, these guys are very normal people. McNair didn't put on any airs. He asked about how we were doing and for details on casualties, all that kind of stuff. Funny how getting wounded can show who you really are.

Battle wounds weren't the only maladies we had to deal with. Simple, even small ailments could become big problems.

Like flea bites.

Toward the end of our campaign in Africa, men started turning up with red dots all over their bodies. They were caused by fleas—the area was infested with them. Goats roamed the countryside, not only in the locals' farms but free, through the woods and underbrush. The problem got so bad that a delousing machine had to be brought up to the front. We'd have whole companies go through a delousing procedure, stripping and then getting treated. Their clothes would be treated separately, with high heat to kill the pests.

It wouldn't have been a pleasant treatment back home. Imagine what it was like under war conditions.

PRISONER OF WAR

As April turned to May, the war in Tunisia turned more and more in our favor. Yet no fight was ever easy.

A key battle was fought at and around Djebel Tahent, a hill known to us as 609. The strategic high spot, which housed positions that were shelling the 16th Infantry, was attacked by the 34th Division without much success. General Allen then sent our regiment's 1st Battalion to take a nearby hill; his idea was that holding that hill would force the Germans to retreat or at least ease pressure on the rest of our units. They took the hill, but the Germans soon counterattacked.

One hundred and fifty prisoners were taken. Among them was my brother Bill.

I didn't know it at the time. It wasn't unusual *not* to see Bill for days and weeks, as our units were not often together. And while soldiers who came to the aid station often brought news, there was so much going on as the division pushed to cut off the retreating Germans that I didn't even know that any of our men had been captured.

Bill and the men taken at Hill 609 didn't stay prisoner for long. Taken to Tunis, they were marched to an old Italian freighter at the waterfront. Along with their commander, Lieutenant Colonel Charles Denholm, they joined about three hundred other American and British POWs aboard the ship May 5. The boat left port early the next morning, only to come under attack from Allied planes.

Bullets ripped through the deck and hull; the ship took on water. The captain turned back. Before he could reach port, fresh attacks scared the ship's crew so fiercely that they abandoned ship, leaving the captain alone on the bridge with the German guards who'd come to watch the prisoners.

The planes kept attacking the ship, which was now dead in the water. That night, some of the British prisoners managed to swim to shore and get help. The air attacks stopped, and the prisoners were rescued, my brother among them. They were all soon back in action, Bill included.

FINAL VICTORY

By May 9, 1943, the Allies controlled Tunis and Bizerte. The 1st Division sat firmly in the Tine Valley; all the roads in the area were ours, and the only thing left to do was accept surrenders and mop up the last diehards.

Free once again, my brother joined me and the other medic staff sergeant at the regiment, our friend Larry Wills, on an informal reconnaissance in one of the villages the 16th Infantry was holding. The Germans had left quickly, abandoning many of their possessions, including a number of vehicles.

One was an Indian motorcycle. Another was an Opel Olympia, a European car made by General Motors. A third was a one-and-a-half-ton troop truck.

My brother wanted the Indian. Just in case it was booby-trapped, we tied a rope to the bike and dragged it about ten feet. When it didn't blow up, my brother hopped on. It started with a single kick.

I took the Opel. It was in good shape considering what it had certainly witnessed in the last few days. There was only one item that needed attention—it lacked a windshield.

A wrecked Jeep a short distance away supplied a replacement.

It wasn't a perfect fit, but with some rope holding it in place, it did the job well enough for me.

Larry got the truck. We drove all three vehicles back to our units, not knowing how long we'd have them, but planning to get the most from them in the meantime.

With Rommel gone and the Germans now completely out of Africa, it wasn't long before we were ordered to return to Algeria. Arrangements were being made for the 1st Division to go back by ship and truck convoy. My brother, Larry, and I had a better idea. We convinced the regimental surgeon, our boss, to let us drive the vehicles back so they'd be available to our units. To our great surprise, he agreed.

And so began a several-hundred-mile road trip through war-torn North Africa. Now that we weren't being shot at, Africa was a beautiful place. The hills of northern Tunis were green with the recent rains; at times the sun would glisten off rock faces as if they were finely cut jewels. The valleys and slopes gave way to a flat expanse of semi-green fields before sliding us to savannah and then desert, then back up to green as we veered in the direction of Algiers at the edge of the Mediterranean. The roads were often dirt or gravel; occasionally we'd make our own paths across flat fields before getting back to hardtop. If it weren't for the reminders of war—broken tanks, busted houses—it would have seemed more like a camping trip than a military maneuver. Not quite a lark, but maybe the closest you could get to one under those circumstances.

Villagers were happy to sell us bread, eggs, and whatever else we needed. Getting gas from the GIs manning fuel dumps was generally easy enough; we'd just pull up and ask them to fill it up.

There was only one exception, about halfway back: a lieutenant in charge of the MPs where we stopped ordered us held until he could figure out whether we were German spies or just crazy Americans. He got on a radio or phone and worked his way far enough up the chain of command to decide it was the latter.

A surprise waited for us in Algiers, the country's capital: WAACs had just arrived and set up camp.

WAACs—Women Auxiliary Army Corps—were American women who had volunteered to work for the army behind the lines, taking jobs like drivers or clerks so more men could have combat roles. For us, the key was that they were women, a species we hadn't had much contact with over the past few months.

We heard about them soon after we pulled in—word travels fast when women are around. They happened to be on their way to the mess hall when we arrived at their base; we asked if we could join them.

"Come on," they said.

The food was great, the company better. A photographer happened to be visiting, and by the end of the meal he had convinced us to pose with the women. Pictures taken, we went back to get ourselves squared away.

Little did we know that those photographs were soon being reprinted in newspapers all across America, telling Americans what a great job the WAACs were doing and making it look like we were all partying over in Africa.

My wife saw it, of course, something she pointed out in a letter some weeks later. Luckily for me, she was the understanding sort.

The WAAC, which dropped one of the *A*'s and became WAC in July 1943, was extremely controversial, with both men and women, civilian and in the service. A lot of people simply didn't like the idea of females being in the army, even if they were far from combat.

But that didn't keep women from volunteering; a total of 150,000 joined during the war.

They did a heck of a job, filling in behind the lines and saving men for combat. They played an important though often unsung role in the war.

TIME TO RAISE HELL

I'll tell you, the 1st Division had quite a reputation. We were first in a lot of things. We fought really hard and we partied pretty hard, too.

I don't know what other units did, but some of our guys would have a beer party before going into battle. It was just the kind of thing you do to get the war off your mind.

So you can imagine what happened when the division returned from Tunisia and settled into camp in Oran.

We'll put those stories under the title of "hell raising." The division certainly lived up to its reputation, something that didn't please the upper echelons of command.

Terry Allen, though, didn't mind.

Personally, I wasn't much of a drinker, and as a married man I wasn't about to be visiting the madams' establishments in town. But as long as they didn't get out of hand, I was fine with letting my guys do what they wanted on their own time. They'd earned it.

Junior officers were in such high demand that the army tried recruiting NCOs to take officer training and get commissioned. I was offered that deal when I got back to Algiers: to become a second lieutenant. Take a commission, join another infantry unit as a combat leader.

I wasn't tempted. For two reasons.

Most important, I liked my unit, I loved my job, and I liked the fact that my brother was in the same regiment. If we'd been close as kids, we were now even closer, even though we spent little time shoulder to shoulder. Being so intense, combat makes all your relationships stronger.

I felt that way about my men as well. I was responsible for them. They were family members. We were working together in a difficult job, one with high risks but potentially great rewards. Saving a fellow soldier's life has a value beyond anything you can measure.

Most of all, taking a promotion would have meant leaving the medical department. At the time, the highest rank you could achieve as a medic if you weren't a doctor was staff sergeant. That changed later on, but in 1943 there was no hint it ever would. As far as I understood, taking the promotion would mean shipping back to the States, leaving my guys, leaving the fight. It would have felt like I'd abandoned my duty and my calling.

There was another reason, too. In the infantry, second lieutenant was just a place to get knocked off right away. The Germans appeared to target the lieutenants, figuring that the platoon would hesitate and lose morale if their leader was gone. That assessment may have been overblown—I knew plenty of second lieutenants who were breathing, not bleeding—but it still looked like one of the worst jobs in the army.

So I stayed where I was. The suggestion that I take a promotion came up a few times again, but I never changed my mind. Staff Sergeant Lambert, Medical Detachment 16th Infantry, was a fine title and job description as far as I was concerned.

HOMEWARD BOUND . . .

One of my jobs as staff sergeant was to censor the outgoing mail.

That one wasn't fun.

I was always careful not to concentrate on what the guys wrote, just to look for what are now called key words, like *Africa,* for example. Anything that might give us away or expose some sort of secret had to be cut out.

That meant all you could really say in a letter was: *I'm okay. I love you and hope you're fine.*

That's about it. Pretty much everyone knew it, too. I don't remember having any problems with the guys over the letters.

Besides regular mail, we had a special type of letter called "V Mail," "Victory Mail," or letters that soldiers and loved ones could choose to send, saving on weight and helping the war effort. After they were censored, the letters were photographed and then sent overseas on microfilm. Written on special paper, the original would be a little smaller than a standard piece of note paper. I'm not sure how big the microfilm was, but you could jam a lot of letters onto the roll. When it arrived, it would be reproduced on a page about half the size it had started with.

Imagine how many people read your thoughts along the way.

At the front, mail delivery was very sporadic. The kitchen trucks usually had a mail bag and would take mail back. It could be months before a letter arrived on either end. It wasn't much better in Algiers, but there was one consolation: after so many months in combat, we figured we were going home.

The rumors started as soon as our troops entered Tunis and Bizerte. First in, First out, pun intended.

In our minds, the war was over for us. Yes, the Germans still controlled Europe, and on the other side of the globe, Japan occupied a good chunk of Asia. But we'd fought what seemed a war and a half. New guys were rolling in; the army was still building up. It was time for someone else to take over the fight.

In my mind, I was as ready to go home as everybody else. I had a son I'd never seen, a wife I barely knew. I didn't want to leave my unit, but I wouldn't be opposed to spending my time with them back in the States, where I could arrange for my family to live with me. And getting shot at every day was never high on anyone's list of pleasures; I could live without that.

Yes, like everyone else in the Big Red One, I was ready to go home.

Uncle Sam, or more specifically Dwight D. Eisenhower, had other plans.

SIX

Husky

GEARING UP

The bazookas gave it away.

Those and the boxes and bundles of fresh gear and medical supplies, along with the replacement troops that flooded into the division toward the end of May and early June. You don't reequip a unit that's going home to rest.

As the rumors faded and reality set in, our regiment shipped out from our camp to Algiers aboard the USS *Thurston,* a troop transport that would serve in both the European and Pacific theaters by the end of the war.

I didn't take the boat. I still had my Olympia, and drove it the couple of hundred miles with the rest of our vehicles to the city. Once there, we started rehearsing for a fresh landing. At that point, we didn't know where our target was, and we still didn't know when we boarded the *Thurston* a few weeks later, weighed down with our combat gear. It wasn't until we were out in the

Mediterranean and Command started passing out pamphlets with crude translations of Italian that it became obvious we were going to Sicily.

Ninety miles north of Africa, the island of Sicily is a bit like an overfilled ice cream cone that fell to the ground and landed off-kilter. It's a rugged, rocky, mountainous island. Mount Etna, an active volcano, is the upside-down ice cream cone; it dominates the northeastern quarter of the island. While there are many beautiful beaches, traveling through much of Sicily means driving or hiking up steep hills and around mountains. The geography greatly limits the ways it can be attacked—though plenty of invaders over the centuries have taken a try at it.

At its closest point, the roughly triangular island is only two miles from the Italian mainland. For the Italians and the Germans, it was a strategic spot that made it easier to defend Italy and project power in the western Mediterranean. For the Allies, Sicily would be a strategic launching point for attacking the rest of Italy. Holding it would give our aircraft and ships bases close to the Italian mainland.

Code named "Husky," the idea of the Allied operation was to land troops in the southeast part of the island. The British would attack up the western side; once again led by Montgomery in the field, the British forces were intended to make the main thrust. Their goal was Messina, the port city at the northeast of the island, close to the mainland. Once the Allies controlled that, any Axis forces would be cut off from the easiest escape route. They would lose the best port to reinforce their army, and most likely would have to leave Sicily completely.

Montgomery was landing four divisions and some additional troops along and through an area about forty miles long in the Gulf of Noto, south of Syracuse. His British Eighth Army was the same hardened, experienced force that had faced Rommel in Africa.

The American forces were to the west. Dubbed the American 7th Army, we were once more led by Patton. The 1st Division was one of three divisions that would hit beaches in the Gulf of Gela, with two units of parachute troopers hitting inland. Our logistics would be a challenge; our landings would be over a fifty-mile stretch, while the parachute assaults were well inland. Aside from countering the German and Italian units in that part of the island, our primary job would be to protect the British flank as they advanced.

If it sounds like we were being given the short end of the stick, we were. The British, Montgomery especially, didn't think we were up to par as fighters. Despite everything we'd been through, we were considered junior partners in the war as soldiers. It was an attitude that, incredibly, would continue through the rest of the conflict.

We'd learned a hell of a lot since landing in Africa nearly a year before. Most of all, we knew how to kill.

My battalion would be part of the first wave, striking near the city of Gela. We expected heavier casualties than we'd had during the Torch landings.

Unfortunately, those expectations would be met.

DEATH IN THE BOAT

I can't remember the exact type of landing craft I climbed down to the night of July 10, 1943. I believe it was a Higgins boat or one of the smaller LCI (Landing Craft Infantry) versions, but honestly I'm not sure.

What I am sure of, what I can still hear, is the sergeant nearby yelling that someone had dropped his gear on the ramp.

And I remember even more vividly realizing that it wasn't a battle pack but a man who had just been killed.

The seas were rough, the clouds thick overhead. The winds were incredible. Forty knots—"Mussolini winds" guys called them, as if the Italian dictator controlled the weather as well as his country. Our little craft had pitched back and forth the entire five hundred yards in to the shore. We'd stopped abruptly, ramp lowered, and as I ran forward I found myself in water nearly waist deep.

Run!

Tracers split through the air around me. Shells were landing in the water, and along the dark shadow of land ahead. Black obstructions seemed to rise from the beach as we got closer. It was a lot like a movie, but one that hadn't been made until then; someone needed to have lived through this hell before such a script could be written.

Ashore, I ran to my right, looking to get my bearings and start helping people. The red cross was back on my helmet and armband; wearing it during the invasion made it easier for the guys to spot me, and we were all targets anyway.

The gunfire and the flashes and the explosions hadn't stopped, but once I was on land I started getting used to them, if that's the right word. As deadly as they were, I'd seen worse. The large ships behind us were answering with their guns. The battalion moved through the dunes, toward the small houses lined up above the beach.

We set up an aid area and got to work. Company medics patched guys up; the bearers grabbed them and brought them down to the open-air station. We got enough of a foothold there for the landing craft to start bringing supplies in.

Then the counterattacks started.

The first hit the Rangers who'd landed in Gela itself, which was to our west, on our left as we came ashore. The crack infantrymen repulsed a surge by Italian foot soldiers, only to be met with a wave of Italian tanks. Though clearly outgunned, the Rangers held on, fighting through the night to keep a hold on the city.

A group of thirty-two light Italian tanks came at our regiment next, heading from Niscemi, a village that was among our immediate objectives a little less than ten miles from shore. The tanks were quickly harassed and picked off by paratroopers who'd landed overnight, then targeted by gunfire from a cruiser offshore, the USS *Boise*.

By the time they reached the 16th Infantry troops, they were down to about twenty. Even so, fighting was intense, made more so by our lack of tanks on the beach. This was a deficiency that would weigh heavily the next day.

In the meantime, the winds that had kicked up the waves overnight punished the small boats and landing craft offshore. A large number were pushed in and beached against their will.

The wind had also pushed our paratroopers off course, sprinkling them miles from their targets. For us, this meant that a key airfield we'd hoped they would take overnight was still in enemy hands.

At least the sun was out. Our battalion secured Highway 115—

contrary to the name, it was a gravel road—and the route to Niscemi by the end of the day. We were off the beach, uphill in the fertile wheat fields between Niscemi and Gela.

By nightfall, the Seventh Army had moved roughly two miles inland. We'd lost fifty-eight men, with another 199 wounded. We'd also taken some four thousand prisoners—the first indication that the Italian troops on the island were not all going to fight as eagerly as we feared.

But there were still plenty who would. As would the Germans.

CRACKING THE CRACK TROOPS

Our aid station was about three hundred yards from the front when the Germans launched a fierce counterattack before dawn the next morning, Sunday, July 11. The brunt of the assault that hit both our regiment and the nearby 26th was once again led by tanks. But this time, they weren't the lighter French-made vehicles the Italians used, but well-armored and heavily gunned Tigers from the elite Hermann Goering Division. The Nazi tanks quickly overwhelmed the companies on the front lines, rushing across the wheat fields as infantrymen either hid or wasted their bullets trying to stop them.

The Hermann Goering Division was named after Field Marshal Hermann Goering, the Luftwaffe or air force chief and one of the most powerful men in Hitler's regime. Though technically part of the Luftwaffe, the division was an armored ground force with over a hundred Tiger tanks; American intelligence rated it among the most dangerous units in the German army.

By 0640, the Germans had overrun the 3rd Battalion of the 26th Infantry and were rolling down both sides of Highway 117. Crashing through farm fields, their guns were now within range of the beach, where General Allen was desperately trying to get some more firepower.

Up near us, in the area of Abbio Priolo, some forty tanks rolled right through our front lines. Two companies broke, retreating pell-mell until the officers managed to rein them in and set up a defensive line at Piano Lupo. A short time later, thirty Panzers hit the 3rd Battalion.

The fighting was so severe that we ran out of bandages, and I had to send back to the beachhead for another doctor to help handle the wounded. We hadn't set up a regular tent at that point because the battle was so fluid; there had been little time to do it overnight and there would be none this Sunday, as we had a steady flow of casualties. By a little after 1000, or 10 A.M., the regiment had lost two battalion commanders and six of the nine antitank guns that we had managed to land.

The situation was beyond critical. The battalion was in danger of being overrun. The beach area was barely secure; Italian dive bombers had made an appearance overhead, and division headquarters was within gun range of enemy tanks. If there was a safe place on the entire island, I didn't know about it.

General Allen, though haggard, maintained his cool. When someone asked if he was going to retreat, he replied, "Hell, no. They haven't overrun our artillery yet."

They were pretty close, though.

Fortunately, the Germans had not been able to coordinate their tank attack with their infantry support. They had some nine thousand foot soldiers—grenadiers and elite troops, many of whom had seen combat in Africa—but confusion and miscommunication had left the infantry behind. That made the tanks vulnerable to attacks from our men as they passed. Soldiers shot bazookas and improvised with sticky bombs attached to the hull of the tanks, knocking them out one at a time.

Since the German vehicles could not stop to properly mop up our infantry, men were able to take cover until they passed; even though outgunned, they were still in place—and alive.

Even so, the fighting was desperate. I couldn't keep up with the number of injuries and had to send back to the beachhead to ask for another doctor. We shifted position to get closer to one of the endangered companies and just kept working, funneling the men back to the regimental aid station as best we could.

Just as it looked as if things would go from desperate to catastrophic, a cannon company and their 105-millimeter howitzers appeared. The big guns fired point blank at the Jerries. Fifteen M4 tanks came up and joined the battle. The *Boise* turned her guns in our direction.

Within minutes—minutes that seemed like days—the battle turned. The Germans began to roll back around 0200.

It was too soon to declare victory. Even as the Germans retreated, dive bombers appeared to blunt our attack. My stretcher bearers worked past their limits, dragging the wounded back to safety. All the world around us seemed to be on fire; what wasn't burning, smoked. A thick fog of burnt carbon and wheat hung over the fields.

The Germans gave up the attack that evening. It was hours before we managed to catch up with the wounded.

Earlier, Patton had decided to reinforce our front by bringing in more paratroopers and dropping them into the area. Unfortunately, that led to a horrific friendly fire incident. Apparently the word that they were coming did not reach everyone; thinking the planes overhead were German—there were still plenty around—antiaircraft guns took aim. Twenty-three transport planes were shot down; another thirty-seven were damaged. We lost about 10 percent of the force in one of the most horrific accidents of the war, certainly on Sicily.

The 1st Division wasn't superhuman; we could be routed, bested, certainly wounded. We did, however, persist and persevere.

One important note: the 2nd Battalion medics never retreated; we just found a better location.

Joking aside, I don't remember us actually moving backward during that engagement, as hectic and desperate as it became.

Going into battle, you know someone is going to die. But you always feel it is going to be someone else.

I dreaded sending the new guys out before they'd had time to get used to battle, and before I had a chance to judge what they could do and how much they could handle. It might take only a few hours to figure that out in battle, but I still felt protective. When you're responsible for men and they get wounded, it can't help but bother you.

I'm the guy who sent him out on that job.

After you've done that a few times, you dread the thought of getting a replacement and sending him out until you get to know him better. You're just more protective of the new guys.

And the old. You have to fight your qualms and your conscience to do your job.

Though battered, the Germans did what they could to slow us down. They blew up the bridges. They posted snipers in the fields, and were very clever about how they laid out and buried their mines. The guys said that iron ore in the island soil made it hard for the mine detectors to work right.

The smoke from the wildfires choked and burned at your throat and eyes. Guys wrapped whatever they could around their faces as masks to filter it.

The Germans abandoned Niscemi and we entered on July 13. The airfield at Biscari was secured the next day after tanks were used to eliminate the last of the snipers.

The proud Panzer division had been defeated, with a number of its soldiers running back in panic at points during the battle. But this victory had come at a terrible cost. I'd lost one of my medics, Angelo Lambiasi. Three others were severely wounded. The battalion itself had 56 dead, 133 wounded, and 57 missing on just July 11 and 12 alone; most of the missing would turn up dead.

We kept going.

The British forces had faced less resistance when they first landed, but things quickly changed. As the German defenses stiffened, Montgomery and his boss General Alexander altered their plans, rearranging the advance and holding our divisions in place on their flank.

The change was frustrating, and it soon led Patton to propose his own alteration, dividing his force in half. While II Corps, which we were part of, would remain on the British wing, he would take the rest on a sweep to the west, circling around the weakly defended island toward Palermo, Sicily's capital.

Alexander okayed the idea—possibly after Patton had already put it in motion. Of course, we had no idea what was going on at that level, and certainly no one came to ask our opinion. All we knew was that we had a hell of a fight ahead of us.

MULES IN THE MOUNTAINS

July 15 found us about fifteen miles north of Niscemi. Looking north, we saw rolling high ground, hills jutting between deep valleys on the mountains. Defenders could use the terrain like a wall, with the extra advantage of seeing an enemy approach long before he was seen.

We caught a bit of rest that day and the next, before preparing to take a key hill a few miles ahead. The plan called for II Corps to fight its way to Troina, an Axis stronghold in the north, above us. By holding it, we would threaten the German army facing Montgomery with encirclement. We'd also protect the British army's flank, with our forces roughly paralleling Montgomery's advance. The British were to move from Augusta to Catania and eventually Messina, a good distance to our right, but they would be vulnerable if the Germans managed to come down through these hills to make their attack.

The plan must have looked good on paper. In the field, the hills and small towns formed so many potential defensive points that it seemed like it had a potential for a very slow and painful slog.

My battalion moved out at 2200 the night of July 22. My aid station was about a half mile from the line, waiting to move up as soon as the enemy retreated.

That didn't happen, not then. The regiment pushed the Germans across a river but then was hit by a heavy counterattack. The units that had crossed the water had to retreat, taking many casualties along the way. Cluster bombs dropped from German attack planes wounded a number of our guys. Others were killed by mines and what turned out to be bombs with delayed-action fuses. These would fall to the ground, often at night, sit there for a few hours, then explode, often while we were passing through the area.

Once more we had a multitude of casualties. We'd restocked our supplies, but it was hard keeping up. Plasma was in short supply. I later estimated my team treated no less than 125 soldiers in a day's time.

The blood we were spilling wasn't in vain. So many Italians were surrendering that our forces had trouble handling them all. But it seemed like every inch of the island had to be fought over several times before we could push on. And it was always uphill.

The roads through the mountains were so poor that there was no way to get trucks up to us as we inched forward. Even getting ammo to the front line became difficult the farther we went. Finally, our regimental commander, George Taylor, hit on the solution—where trucks couldn't go, mules might.

The way I heard the story, he found a local farmer who had a hundred mules. The farmer was willing to sell the mules to him, as long as he got paid in cash.

No money, no mules. And certainly no IOUs.

Taylor, of course, didn't have enough pocket change to cover the cost. He sent back to division headquarters, and they gave him the money.

Probably after a bit of head scratching when he told them what it was for.

The animals clambered over the rocky paths, living up to their reputation for stubbornness, but they did ease our supply problems somewhat.

They also gave at least one of our city-slicker GIs a good toss after he tried to play cowboy and ride one. That ended his cowboy days, I suspect.

At least that's the way I heard it. The story of his ill-fated rodeo shot through the ranks like a hurricane whipping over a barrier island. Good stories and wild rumors always do. I'm sure by the time we left Sicily, the tale had him being thrown all the way to Mount Etna.

There's probably more truth to the stories about the mules only understanding Italian, which meant locals had to be enlisted to help drive them. Then again, mules are stubborn in any language. There was only one mule-related injury that I can personally vouch for—one day a guy came in with a broken foot. A mule had stepped on it.

I'd guess that soldier wasn't a farm boy, either.

As the companies climbed through the mountainsides, we would move up behind them, generally sticking close to the roads running through the valleys. It was easier to bring a wounded man down by stretcher and evacuate him from there if necessary than to locate up on the hillside. As always, we tried to locate the aid station in a place out of direct fire. The German air force was more powerful here, and besides their artillery and mortars, we were vulnerable to bombings and strafing attacks. But mortars remained the most dangerous weapons. Since they were highly mobile, not only was it difficult to take out the mortarmen, but an attack might come at any time.

On July 24, mortar rounds hit the station where I was working just outside a small village in the area south of Troina. I caught some fragments in my back. It hurt, but it wasn't bad enough to go back to the hospital. One of the doctors picked out shards with tweezers; one was fairly long.

"Sergeant," said the doc, "now we'll give you a day off."

"Thanks." I laughed. "And where will I take it?"

Sulfa, a bandage—then I went right back to work. I guess the army still owes me that day.

Around the time that mortar shell tattooed my back, one of my medics came back from his company with a stomach flu. I had him rest at the station and decided I'd fill in for him the rest of the day.

I went up and saw the company commander.

"Sergeant, what are you doing here?" he asked.

"I'm taking the place of my guy."

"Great. Stick close to me."

"Are you trying to get me shot?" I answered.

I meant it as a joke, naturally, but within no time at all things suddenly got hot. If I'd had any doubts about how tough a job the company aid men had, they vanished with the first gunshots. It was a struggle to stay visible as the firefight continued. I hid behind a rock when the bullets got to be too much and found little bits of shelter as I jogged around the positions, but mostly I was out in the open.

The fighting was heavy. Things got so bad at one point that the commander told his men to get their bayonets ready. The Germans were so close he feared they would rush us.

"Fix bayonets!"

I didn't like the idea of that at all, and not just because knife wounds can go very deep and cause a lot of bleeding.

Fortunately, it didn't come to that. The Germans finally decided they'd had enough and slipped back. The shooting died down, and by nightfall it was quiet again.

Amazingly, only one GI was hurt the whole six or so hours I was there. It wasn't even a bullet wound: his helmet strap had somehow managed to wear into his skin and cut it badly enough to bleed.

He was the only guy I ever got to treat as a company man.

INTO A BURNING CALDRON

We were moving on the west side of Mount Etna a few days later when Dr. Morchan and I took a Jeep out to check on our company men and find a fresh spot for the aid station. The battalion was spread out in the hills, and we found ourselves riding down a narrow road behind a pair of American M3 tanks. They were not going very fast; we were so close to the second tank that I could feel the heat of the exhaust in my face as we came up the road. If it had been winter, that would have been welcome, but by now it was fairly warm, and the dust and noxious fumes were annoying.

I dropped back a bit, slowing down and trying to decide whether and where I might be able to pass. Then I noticed flashes and explosions ahead. It was hard to tell from the road, but my guess was that an American position was being shelled from across the valley we were in. The tanks must have been going to reinforce whoever was being attacked.

Before the doctor or I could say anything, something hit the tank.

"Don't!" yelled Dr. Morchan as I hit the brakes. "Don't!"

I leapt out from behind the wheel and began racing for the tank. Two of the four men crewing it managed to get out, stumbling away from the smoke and small flames inside.

"Lambert, this is a direct order! Do not go into that tank!"

I don't know if those are the actual words he said, but Captain Morchan gave me a direct order not to go into the tank, which was smoldering. He might just as well have told me to stop breathing.

I ran past the two tankers who'd gotten out and grabbed one of the handholds to pull myself up. Atop the tank, I reached into the open hatch on top of the turret and crawled inside.

"Lambert! No! Ray! I'm ordering you back! The tank is going to explode!"

I managed to get enough of a hold on one of the guys to pull him through the turret. I helped him to the ground, then leapt back onto the body of the tank, crawling inside to get the last man.

Somewhere along the way, my hands started burning. Thc sensation was distant, far enough away that all I felt was the weight of the crewman as I prodded and pushed him upward into the hatchway, then through and out. After we stumbled to the ground, I started dragging him away.

My field jacket had caught fire while I was inside the vehicle. I patted it out, then moved everyone back about forty yards as the flames shot out of the turret and the ammunition started to cook off, exploding with the heat.

Dr. Morchan was beyond angry with me, threatening a court-martial. The two tankers I pulled out lived, as did the others; maybe that was why he didn't go through with it.

Was it a brave thing or a stupid thing to do?

Both, I guess.

TROINA

The tanks had been heading toward Troina, our next objective. The Sicilian city sits on a hill of rock, surrounded by boulders, perched in the mountains. It's the highest large settlement on the island, which was one of the things that would make it so hard to take.

Patton had seized a virtually undefended Palermo and was now moving eastward along the top of the island. The British were making progress on the east coast of the island—slow progress, but they were moving up. Troina was a key city between the two forces. It was at the head of a valley that ran along the northern side of Etna, giving it access to both the north and east.

General Bradley, the II Corps commander, moved our division and the 45th into position to take the city. Thinking the city was not well defended, General Allen originally had the 39th Infantry Regiment from the 45th Division attack August 1. They were thrown back, and it became obvious that there were a lot more Germans in and around the town than we'd thought.

The 16th Infantry began its attack on August 3. We met heavy resistance right away.

The Germans took full advantage of each nook and cranny on the paths into the city. They had machine guns set up behind just about every big boulder possible.

The terrain made it hard for us to get the aid station close to the front line. The hills made for irregular battle lines, which left us vulnerable to attacks from three if not four sides, rather than the usual one. I set up originally between a quarter and a half mile from the companies near the road below most of the fighting. Our guys were fighting on terrain practically 90 degrees straight up a short distance away. You could hear the gunfire and weird echoes of the explosions. We had a constant flow of men in and out of the station.

When the troops finally started to see some progress, I decided that I wanted to move the aid station up. I felt we needed to lessen the time it took to get the men in, not only so they could be treated more quickly, but so they wouldn't be exposed for so long a time on the way to the station. The wear on my stretcher bearers would be less as well.

I took the Jeep and a driver up the road—if you want to call it that—to see where we might set up. As we turned around a corner and headed down into a narrow valley, I saw a soldier lying on the ground who'd been hit by a mine.

I hopped out to get him, despite my driver's complaint that he would do it.

"I got him!" I yelled.

It was just like back in Africa. Déjà vu all over again.

At least I hoped so.

Fortunately, this soldier was very close to the road, because there wasn't much of a track to follow. I figured the most likely path, got to him, pulled him up, and carried him back to the Jeep.

There were probably more mines along the road; I was just lucky. The Germans wanted us to bleed for every inch.

And we did.

The aid station stayed where it was. Early in the afternoon, the second battalion commander reported that E Company had lost contact with two of its platoons. F Company had been so devastated by the attack that only one of its platoons was still at fighting strength.

Some of those missing soldiers were at our aid station, getting patched up. The rest were dead, lying where they'd fallen, stranded by German gunfire and counterattacks.

We were overwhelmed, barely able to see the wounded before sending them down the road to safety. Once more we were running short of supplies—it seemed to be a constant theme in Sicily.

Both sides poured artillery back and forth for hours. Our planes bombed the city. In some places our companies were fighting almost hand to hand.

The constant flow of casualties continued. Big shrapnel wounds that earlier in Africa might have seemed serious were now routine. Chest-sucking wounds that threatened to drown a man in his own blood, gut wounds that exposed the stomach and colon and everything else—those were the evil ones. The doctors worked furiously, mechanically, continually.

So many men.

Just from the flow of casualties and the sound pounding in our ears, we knew the battle wasn't going well. By 1635—4:35 P.M.—the battalion was pulled out of the fight. We were ordered to get ready to withdraw.

We'd made a dent, but that was all. The unit had been severely mauled. The attack would continue, but without us.

TERRY SACKED

II Corps captured Troina on August 6, after what turned out to be one of the toughest battles of the war. It took the 1st Division, the 9th Division, a French Moroccan infantry battalion, over 150 artillery cannons, including massive 155s, and countless aircraft to finally dislodge the Germans.

The first guys in found over a hundred bodies on the streets.

The center of town looked like a rock quarry with all the rubble from the destroyed buildings. Flies buzzed over the dead; maggots covered the corpses.

The 1st Division had over five hundred casualties in that battle. Those of us who hadn't been hit were exhausted.

We'd been going nonstop since landing almost a month before. For the medics, that meant snatching sleep at odd moments when we could, often for no more than an hour or two. A typical day might start when you woke from a nap a few hours before dawn. You'd mix up some coffee from the instant stuff with the C-rats. The first casualty would come in while you gulped cold pork and beans. You'd check the bandages and give the guy a blanket against the cold, hoping to stop his shivering.

You'd be shivering yourself in the cold mountain air. By noon, you'd be sweating. You wouldn't hear the mortars anymore; they were just background music as the walking wounded flooded in. A man would be brought in and laid on the operating table; even before the doctor came over, you'd know the result wasn't going to be good.

We'd work on him anyway, desperately.

Jeeps would roar off every so often, stacked with bodies fore and aft, racing to a place where the men could get further care, or maybe find a more peaceful corner to die. Night came, but you wouldn't notice; the only way to keep track of time was by the ebb of wounded, the flood turning to a trickle even as the mortars kept pounding the hill above. When there was nothing left to do, you'd grab a nap, only to be woken again far too soon.

We needed a rest.

We were able to get some in the days that followed, as other units moved ahead to Troina and then beyond.

Having lost so many men, our morale was already sliding when we heard the news that Terry Allen and Ted Roosevelt had been fired.

Depression quickly turned to anger. How could they fire Fighting Terry Allen? Or mistreat Teddy Roosevelt, son of a president?

All sorts of theories have been hatched and different generals blamed, but we knew at the time it was Patton. He didn't like being upstaged, and it was clear anytime he was around Terry Allen that Allen's men liked him a hell of a lot more than they liked Patton. We might fight for Patton, but we'd go through hell and back ten times again for Terry. And then we'd fight.

Patton had never liked General Allen. There's a famous story about one of their first meetings in Africa. Patton had recently taken over and was inspecting the division headquarters. He was shown some trenches near the command post and told that the officers would take cover there when the German planes attacked.

He asked General Allen which of the trenches was his. When Terry pointed it out, Patton went over and peed into it.

It's not recorded whether the urine spelled out the word "coward," but that was clearly implied.

There's a line between a coward and a complete fool; standing unarmed and unprotected as planes are dropping bombs on you is way over the line into fool territory. Terry Allen was no fool, and certainly no coward. He led from the front, often exposing himself to fire, and the men loved him for that.

It wasn't just courage that made him great. He was aggressive, yet the men felt that he cared about them as well. Terry could sit down next to you on the ground and just ask how you were. He didn't make a big deal out of clothing regulations or other trivial baloney. He knew that guys needed to blow off steam, and he let them. He enjoyed a good drink, and he didn't mind if his men did too. He wanted results, and he felt that to get them he had to give us a little leeway.

I'm not saying he was a perfect general, let alone a perfect man. But he was a great leader, and his firing wounded us deeply.

Generals aren't actually fired in the army; they're "reassigned." But there's no mistaking what happened here, especially since it happened right after a difficult battle.

Those sorts of reassignments were traditionally a prelude to retirement. Disgrace is implied, if not spoken aloud. But Terry Allen always made his own path in the army, and he was neither a quitter nor a man who would run from a challenge. After he returned to the States, Eisenhower appointed him commander of the 104th Infantry Division. His "Timberwolves" went to Europe the following year, joining in the final drive against the Nazis and winning praise from the men who had reassigned him in '43.

Ted Roosevelt, the general who was Terry Allen's deputy at 1st Division, became the assistant commander for the 4th Infantry Division. He, too, would be heard from again. Destiny was biding its time for him on Utah Beach in Normandy.

There are always odd stories you hear in combat, tales that are even stranger than wild rumors. Because in war, things happen that you could never make up.

One of those involved some paratroopers who were captured shortly after jumping the first night of Husky.

The wind took aim at their round canopies as they went out of the plane, pushing them all over the place, well off target and into some very rough landings. Several were hurt, and German patrols had no trouble rounding up a half dozen or so. They took the men to a villa in the mountains, presumably to await further orders.

But things got heated. As the fighting escalated during that day, the German infantrymen also sustained casualties, who were also taken to the villa. By the time night fell, the villa was close to the American lines.

The lieutenant in charge of the German detail decided to offer his American counterpart a deal. If the GI, also a lieutenant, agreed to get care for the German wounded, he would let him and the rest of the paratroopers go.

It seemed like too good an offer to pass up, even if it was from a man whose job it was to kill him. The American lieutenant traveled down through the lines and arrived at the aid station run by my good friend 1st Battalion Staff Sergeant Larry Wills. After he explained the situation, several medics volunteered to go back with him in Jeeps to take out the wounded Americans. They'd also patch up the Germans before leaving.

I can imagine the tension they must have felt on that ride up. What I can't quite picture is the look on their faces when they found the soldiers from both armies sharing wine from the cellar.

And, supposedly, singing.

The German officer apparently not only upheld his end of the bargain; he gave the Americans the password for their night patrol in the area. Everyone got out okay, including the Germans.

What happened the next day or the next, no one could say. But for that brief moment, the men on both sides were just men, not mortal enemies.

One other thing I remember about Sicily: Mount Etna. The volcano, which towered high above us, shaded the valley and surrounding slopes a strange red. It was as if the planet was having its own war below.

Major General Clarence R. Huebner took over for General Allen. We didn't particularly like him. We thought maybe he came from a Girl Scout camp, the way he fussed over trivial things like the proper form for salutes and uniform regulations. Where Allen cared less about the b.s. of army life, Huebner was a stickler for it. When it came to discipline, he was practically Terry's opposite—and you wouldn't think of calling him Clarence, either, not even behind his back.

The fact that he was such a stickler was undoubtedly one of

the reasons Omar Bradley wanted him; he felt the Big Red One had become undisciplined, and that was hamstringing our performance in combat. Bradley was the corps commander and would have had at least some say in the appointment, though Patton and Eisenhower would have had more clout, and ultimately George Marshall was the boss.

Discipline is good; no organization, no army, can succeed without it. But you can go overboard, and Huebner did. He was so much stricter that it bothered the guys, and he didn't need to be that way. We'd been through two hellacious campaigns by the time he joined us, but in our eyes he treated us practically like newbies, wanting us to start from scratch. Even if he hadn't replaced a beloved general, he would have had a tough time with us.

Give him his due, though. Huebner started in the army as a private and worked his way up the ranks. Only a handful of men have ever done that. And twice he won the Distinguished Service Cross. He was a 1st Division lieutenant colonel in May 1918, in the battle at Cantigny.

"For three days Lieutenant Colonel Huebner withstood German assaults under intense bombardment, heroically exposing himself to fire constantly in order to command his battalion effectively," reads his DSC commendation, "and although his command lost half its officers and 30 per cent of its men, he held his position and prevented a break in the line at that point."

Two months later, in mid-July 1918, after all of the battalion's officers were killed or wounded, Huebner led his men and those from another battalion in an advance against the Germans. Though wounded, he managed to achieve his objective, and earned an oak cluster for his Distinguished Service Cross. (The cluster is pinned on the ribbon, and is the army's way of saying the award was earned a second time.) He also won a Distinguished Service Medal and a Silver Star in the war.

Huebner did eventually earn the respect of the division's soldiers, but it took a long time for the guys to come around to him. And he was never liked anywhere near as much as Terry.

PATTON

As much as we all hated him for sacking Terry Allen, I have to admit that Patton was a brilliant man.

He had a certain way he liked to do things, and he talked a different way than other generals. But when he went into battle, he was very intent. He was good at war.

Of course you've heard all about George Patton, how he walked around with a swagger, had pearl-handled pistols, was all blood and guts. You may not have heard that he composed poetry in his journal, or that he was deeply in love with his wife. He couldn't stand British general Bernard Montgomery, though that can be said for just about every American general, with the notable and important exception of Eisenhower and his staff.

Having seen Patton up close, I'd say his flamboyance came from feeling inferior growing up. I think he felt he had to constantly prove himself when he was a child, and never got past that.

There's also the possibility that he was dropped on his head as a kid. He really did have his moments of being crazy.

His ego didn't make him any less brilliant, but it did make him less well liked. And abrasive.

At the same time, it made him memorable. We don't seem to value the quieter men anymore—people like Huebner, I suppose, or Bradley, who tended to go about their business much more quietly than Patton did on a bad day. There were a lot of those guys in the army back then. Unfortunately, a lot of what they did has been forgotten by history . . . but not by the men who served under them.

Who will remember them when we are gone?

Among the things you've heard about Patton, I'm sure, are the famous, or infamous, slapping incidents.

Two incidents took place while we were on Sicily. The bare outlines are these: Patton visited wounded troops in field hospitals at the beginning of August while touring our II Corps areas. At Nicosia, on Cyprus, he went to the 15th Evacuation Hospital. There he is said to have slapped a soldier who was not visibly wounded and told the doctors to send him back to the front. About a week later, he did something similar to another soldier at the 93rd Evacuation Hospital.

Eisenhower eventually heard about the incidents, though apparently not through official channels. While he scolded Patton for them, he didn't directly punish him. Those incidents are often blamed for Eisenhower's decision to put Bradley rather than Patton in charge of the D-Day landings, but most historians say it's likely that the decision had already been made for other reasons.

We heard about the Nicosia incident ourselves within a few days, if not hours, of it happening. That was the field hospital where we were sending our wounded, and we had plenty of communication with the guys there.

The more lines of communication you have, the more information you get—and sometimes that obscures rather than illuminates the truth. That may have happened in this case, where second- and third-hand accounts probably mixed with firsthand descriptions and personal prejudices. The version I believe is closest to the truth goes like this:

Patton was visiting the wounded recovering in a ward tent when a young soldier started shaking uncontrollably, unable to stop. The guy started laughing and crying and everything else and just went to pieces. Patton went to him and slapped him on the cheek and said, "Right! It's time to come out of that now." Then he told the GI to get back up to the line, there was nothing wrong with him.

In other words, the general was simply trying to snap the soldier out of a hysterical fit, telling him he was okay, and getting him back to where Patton thought he belonged, rather than sending him back to a hospital and what surely would have seemed disgrace to many at the time.

Now, we didn't know that much about post-traumatic stress disorder (PTSD) at the time, but the man would seem to have had it. We called the condition battle fatigue, or used an older World War I term, "shell shock." Or maybe "nervous breakdown." Officially, the doctors classified it as psychoneurosis, a vaguely defined mental condition; anyone diagnosed with it was not considered a battle casualty, even if it was obvious to us that being in battle had caused it.

The most popular treatment for it, and one approved in the medical department, was to do roughly what Patton did, without the slap—tell the soldier to get hold of himself, let him have a brief rest of a day or two, and then send him back to duty.

I don't like Patton, but I think he gets a bum deal on that.

As in North Africa, the aid station was a great place to meet generals and other VIPs; it was close to the front without being under direct fire.

Usually.

It also featured what passed for creature comforts in the field. Coffee, which came out of a C-Ration and was roughly equivalent to instant coffee today. And alcohol.

The alcohol was for sterilizing needles and instruments. But more than one GI used it to spike orange juice, which was available from time to time when the kitchen trucks came up, or occasionally at the officers' mess. A man would save the drink, then add a little something when he came over to see how we were doing.

How many generals and VIPs drank the field version of a screwdriver?

Many.

I won't name names, but you'd be surprised at a few.

MALARIA, TRENCH FOOT, AND OTHER DISEASES

The VIPs could be a welcome break or a nuisance, depending on the situation and the individual. The civilians we treated were usually stoic tragedies.

We didn't get many because we were so close to the fighting;

the Sicilians tended to flee well before battle and wait until an area had been secured for a few days before returning. But we did get some. I remember treating a man who came with his foot half blown off; he'd stepped on a mine planted by the Germans in a farm field.

Less dramatic were the malaria cases. Mosquitos infested every bit of water on the island, from puddles to deep wells. The females passed the parasite that caused malaria to humans whenever they fed. It could take two weeks for the symptoms to appear. In the early stages, victims might easily mistake them for a mild flu, and continue to wear down their body as the parasites slowly blew up their red blood cells, causing anemia and liver ailments.

If malaria was a problem for the civilians, sapping their strength, it was far worse for the soldiers, and we had an epidemic well before we left the island. Guys fought on the front line with fevers, barely able to walk down to the aid station when the action lulled. They'd be shivering and dead on their feet after a hundred yards. The worst would have yellow faces, caused by jaundice and anemia, as the parasite destroyed red blood cells.

Making a diagnosis in those cases was easy. Treatment was a lot harder. We could give soldiers quinine, whether the real stuff or Atabrine, an effective but horrible-tasting synthetic drug substituted because of the war, but prescribing rest was impossible. And unfortunately, rest was what they needed most.

Trench foot was another huge problem, though this affected GIs, not civilians. A soldier's feet would swell, turn purple or black, and become numb. In a very severe case, blood would be cut off to the toes. Without blood, eventually the entire toe would die—gangrene.

Trench foot was caused by the feet never getting a chance to dry after being soaked in water. It was common in World War I, where soldiers spent days and even weeks in water-filled trenches with leaky boots or shoes. In our case, the guys started off in Sicily by running through water. Their socks and boots never got a chance to dry out during the first thirty-six hours after landing. There was just so much action going on that there wasn't a chance to take off their boots or shoes, air everything out, and switch to dry socks and, ideally, another pair of boots or shoes.

Pretty much no one had two pairs of shoes with them, nor was there time to let the ones they had dry out. Everyone had been issued two pairs of socks, with the idea that one pair would be washed and left to dry while the other was being worn. That was fantasy for most of us. Even when guys had a chance to wash their socks, there was little opportunity to let them dry.

In many cases, the damage to their feet had already been done in the first day and a half. It just got worse as they hiked. I remember treating fifty cases at our station during the first week on the island. These were extremely serious cases, with men's feet so blistered and swollen that they couldn't walk. In some cases, they couldn't even get their boots on or off.

We didn't have drying agents with us, let alone proper antifungal medicine. So we had to figure something else out. We mixed a concoction of aspirin and alcohol into a paste, then rubbed it on the infected toes and feet. The aspirin and alcohol killed and dried out the fungus. We had clean socks brought in and gave them out to the men we treated. They needed no reminders from us about how important it was to keep their feet dry for the rest of the campaign.

Minor cuts and scrapes were another matter. Many soldiers would complain that they just weren't healing. There could be any number of reasons, but a number blamed the island itself. "Sicilian Disease," they called it, as if the rocks could literally suck life from them.

In many cases, the soldiers were just too busy fighting to be careful with their wounds. They were impatient to be healed, but also weren't about to fuss over a little dirt, or make sure a wound was aired and redressed properly.

Medical supplies were often short, especially during the high points of battles. The problem extended to the hospitals as well. And even when the fighting slacked off, we sometimes couldn't get enough ambulances for transport.

But we dealt with it, patching up guys, sending them back or moving them along, taking care of the next one in what seemed like an unending treadmill of hurt and heal.

VICTORY

By the time Troina fell, the Germans must have realized that there was no stopping us on Sicily. They began a full-scale retreat to the mainland. Meanwhile, Patton pressed east along the north from Palermo, battering the rear guard of German units determined to save themselves for a different fight. Montgomery continued to move up the eastern side of the island, making better progress each day. II Corps, our parent group, pressed through the mountains toward Messina.

By August 17, the Allies controlled Messina, and resistance on the island had all but ended. The tenacious fighting in the mountains had bought the Germans time to get the bulk of their forces over to the mainland; Allied soldiers would pay for that in the future.

We were regrouping in a small town near Augusta shortly after Sicily had been liberated when word passed through the unit that Eisenhower and Bradley were coming to inspect us.

This wasn't a normal visit, nor were they checking to see if we were having trouble with our stocks of bandages or plasma. The generals were coming to hand out medals. The cameras rolled while Ike pinned a handful on the officers.

General Bradley handled mine and a few others', pinning a second Silver Star on me, awarded because of my stunt in the Sicilian minefield.

When we were done, we all went back to work, waiting for orders we were sure would send us home.

SEVEN

(Some) Rest for the Weary

HOME, SWEET HOME . . .

I never got the paperwork for that medal, but at that point I didn't much care; I figured it would come in due course.

Besides, the war was over for us. We'd done our jobs, fighting for the better part of a year. We'd been in action for weeks and even months at a stretch. We'd done our duty and earned our passage back home.

Maybe a ticker-tape parade? Or at least a small party with the family.

That's what we were thinking as we boarded the ship in Algiers and headed toward the Atlantic that October, two months later. Home was calling, and our heads were filled with so many sweet memories and grand visions that it was surprising the vessel didn't just sink from the weight. I had a wife to get reacquainted with and a baby to meet. I wanted to see my parents.

I wanted to sleep. I wanted not to hear noise. I wanted to close my eyes and not worry about how long I dozed off for, or worry over what new replacement I had to send to dodge bullets for the first time.

I wanted peace.

Trudging up the gangplank and onto the deck, those dreams seemed only a few days away.

Then we realized there were English civilians aboard our ship.

Slowly it dawned on us that we weren't going home at all. Instead, we were going back to England to refit and retrain.

The war was far from over, especially for us. We were the Big Red One, the army's most experienced, most accomplished infantry division. We were too precious to be rotated home, let alone released to return to our families. We were the Big Red One, and it was our job to end the war.

I spent most of my days in the medical station above deck; it wasn't the lap of luxury, but it was a lot better than the crowded bays and passages below. We were just busy enough for our thoughts to stay focused on what we were doing; the work was just easy enough for us to feel as if we were having a rest. After tourniqueting thighs amputated by shell fragments and patching groins mangled by Bouncing Betties, the worst compound fracture seemed like an easy fix. Being able to dispense aspirin for a fever without worrying about when the next supply might arrive was a treat rivaling the sweetest pie.

We landed in Liverpool on November 5, 1943, and headed to Bridport, a quaint collection of villages not far from the English

Channel. The area had once been known for producing rope. In the days before the war it was a market center, a place where fishermen might sell their catches wholesale and retail. It now had a more urgent task: housing soldiers for the coming invasion of France.

Though we'd been away for barely a year, England this time around was a different place. There were still shortages and a great deal of sacrifice; those were constants throughout the war. Fortunately, bombings had become far less frequent; and while people remained uneasy, the planes that flew overhead now seemed always to belong to us.

There was no sense that the war was going to end quickly, but there was a feeling of progress. Where once the newspapers had trumpeted stories of desperate commando raids as morale boosters, now the pages were filled with more honest appraisals of actions on far larger and more meaningful scales. Africa had fallen, then Sicily. Italy had changed sides and become a battleground. The Russians were fighting back.

But in a sense, we were the biggest difference. There were more Americans in the country than when we left. Far more. And we weren't junior partners in the war any longer. While our experience didn't match that of the British—we had been fighting for less than half as long as they had—we'd learned many of the same lessons, and paid out tuition in blood.

Some of the British officers at the top didn't or couldn't accept this. There would be friction between commanders like Montgomery and Patton for the rest of the war. But a lot of the "Tommies"—the regular British soldiers—did. Most of the ones I dealt with accepted us as equals.

Or close. You know how soldiers are; their unit is always the best.

The military vehicles that plied the streets of Bridport and the soldiers who filled the pubs made it impossible to think life was anything like normal in southeast England. German soldiers were building bunkers barely eighty miles away. But life went on. Bits of "normal" poked through the martial façade like dandelions poking through the pavement. Shopkeepers washed down their thresholds every morning; locals sifted into the pubs in the evening. England carried on.

Our regiment was spread out through the area around Bridport. Some of the line companies had to settle for tents, but my med team was billeted in a pair of houses about a block from the center of town. We set up the aid station in one of the houses, taking over the downstairs living room as our examining/operating room; my office and bedroom were just upstairs.

Our work was light, a world away from what we'd done on Sicily. I would always have someone on duty during the day just in case, but the most serious ailment for weeks on end was a cold. Not that any of us complained.

A dentist came in and we all became dental assistants. The drills the dentist used were operated by foot, the way sewing machines were. We'd have two fellows take turns, cycling the pedals while the dentist drilled.

I went as easy as I could on the guys, not only because of what they'd gone through but because I had a notion of what was ahead. I didn't make a fuss about their sleeping late. I gave them every break I could. I knew they'd appreciate it, and would pay it back when we needed it most.

There were a few diversions. My brother Bill was billeted a few miles away in Walditch, and we got together whenever we could. I remember trucks regularly taking guys to London. Looking for something different to do, Bill and I went to a tennis match and watched for hours. We didn't know what the devil was going on, but it was different.

There were dances in the city, and big bands filling clubs with very good music. And to me, all the old things were interesting—castles and buildings from the Middle Ages, remnants of wars many centuries gone.

I made friends with a local baker, who not only kept us well supplied with bread but occasionally invited me to his home for dinner. He had two daughters, one of whom was in the British army. A lot of the guys were dating English girls. These weren't just wartime flings; I believe the division had eighty marriages come out of our stay there.

Yet looking back, the sky always seemed gray for us that winter. We were resting but not quite relaxing. We were not home, and most of us knew there was no chance of going home until we went through one more stretch of hell. And as bad as things might have been in Tunisia and Sicily, it would be far worse when we hit the shore eighty miles away.

We took our meals in a temporary mess hall set up in a building roughly across the street from that house. For Thanksgiving and Christmas we got special rations—turkey with the trimmings. Otherwise the food was filling, about on par for army food—better I'm sure than what the British were getting, but not something that would make a gourmet lick his lips.

We were down the block from an all-girls' school, and in the afternoons some of the guys would go out on the street to watch them pass. It made me realize how young some of our replacement soldiers were.

And how lonely.

THE HOME FRONT

Back home, people hunkered down for another winter of war. Certain things that would have seemed odd two years before were now routine—blackout curtains or shades on windows on the coasts, cars without headlights at night. With imports severely limited and production shifting to items the army needed, rationing had been instituted to allow for fairer distribution and eliminate price gouging, or at least send it underground.

The government sent families ration books with stamps, which had to be used to buy items like meat, sugar, and cooking oils. There were strict rules governing the stamps; if someone died, his or her stamps had to be turned in and couldn't be used by another member of the family.

There were also strict price controls on many items. People were urged to recycle tin cans and even fat for the war effort. Posters encouraged people on the "home front" to do their bit by planting victory gardens and canning vegetables and fruit to lessen the need back home.

And of course this was the time of Rosie the Riveter—women took on factory and other jobs to make up for labor shortages caused by increased production and the need for soldiers.

Movies were a big entertainment and diversion. A number were war movies, fictionalized versions of what we and our brother servicemen were going through, like *Guadalcanal Diary, Destination Tokyo,* and *The Life and Death of Colonel Blimp.* Laurel and Hardy, a famous comedy team of the era, managed to make air raids look like fun in *Air Raid Wardens.* There were family movies like *Lassie Come Home* and horror stories like *Day of Wrath* and *Son of Dracula.* Serial heroes soldiered up for the war in movies like *Tarzan Triumphs.*

Most of these films are remembered now only by historians or movie buffs, but they were part of a shared experience back home. People weren't just supporting soldiers; they were thinking about and praying for their husbands, their sons, their brothers. The boy who delivered your newspaper, the doctor who had looked after your measles, were now getting shot at on the other side of the world. Without real news beyond what might be gleaned from a V letter six weeks old, people hungered for whatever assurance they could get, no matter how vague or unconnected to the actual war.

It must have worked that way for us overseas, too.

A lot of the popular Christmas songs playing on the radio and selling in the record stores that year celebrated the good old days and family. "I'll Be Home for Christmas" was a huge hit for Bing Crosby. Norman Rockwell's famous painting *Freedom from Want,* showing a family sitting down to a turkey dinner, was actually painted for a magazine cover earlier in the year, but it caught what everyone in the U.S. was wanting—their "boys" home safe and sound.

When would that be?

Soon, maybe. But first there would be more war. Christmas Eve fell on a Friday that year. That afternoon, Franklin Delano Roosevelt went on the radio.

"Our Christmas celebrations have been darkened," said the

president. "We have said Merry Christmas and Happy New Year, but we have known in our hearts that the clouds which have hung over our world have prevented us from saying it with full sincerity."

That would change soon, he predicted. The coming year would see the Allies push ever closer to the root of the evil.

An invasion was planned. Eisenhower would lead it. And the only possible result would be complete and total victory.

MY GUYS

War ages you up real fast. Our team members were in their early twenties, but they had the life experiences and maturity of far older men.

The core of the team had been together since before Pearl Harbor. We'd lost some good men, but those who remained made our unit the best in the medical department.

My opinion, of course.

The faces roll through my mind as I look back, names slipping off my tongue . . .

I remember John Givens as a truly dependable guy, an older man, or at least older than most of the rest—"old" is such a relative term, especially to me now.

Herbert Glassford. Stocky and strong, exactly what you need in a litter bearer.

Doyle Helms. Shy. Liked to sleep in his own tent, away from other guys.

Stanley Appleby. One of my stars. You couldn't find a better company man—though we had equals in our group.

Leroy Kisker. Pale. A whiz at fixing things. Good with the tech stuff.

Alton Pitt. Tall enough to be a basketball player. My man in charge of the new litter bearers.

Ray Lepore. Tech 5, a little older than some of the other guys. Came on with us in '42 in time for the invasion of Africa. Football player back in the Boston area. A fine soldier who caught on to new things immediately.

Herbert Meyers. Had to be the easiest-going guy in the unit. Great demeanor. I made him a company man because he could get along with anyone and everyone. He was a Tech 3, senior guy, a leader.

And then there were the new replacements, men who came in to us, filling in for the guys we'd lost along the way.

Charles Shay was one of these. Quieter side. He had a different background than most of us—he was a Penobscot Indian.

I had no say in who I got. I told headquarters how many men we needed, and they would send men over. For the most part, the guys who joined us had not had any training, at least not to the level we needed. We trained them ourselves, teaching them the fine points of bandaging while mortars are exploding nearby, showing them how to do a transfusion when the night is pitch black and there's no one there to hold the flashlight, and all the other stuff they needed to know to help in combat.

We went out on marches, building up endurance and strength. Fifty miles with the battalion, twenty-five out, twenty-five back, camping overnight along the way. During training, the company aid men would work through maneuvers with their company, so they'd get to know not just the procedures but the men, and vice versa.

I have a photo taken before the war started, and I'm the only one still alive. I'm the only one still here to remember it.

Time humbles all.

My brother Bill was no longer with the medics.

Following the battle of Troina, Captain Joe Dawson had been assigned to take over G Company, whose leader was shifted to another company in the battalion. Dawson asked my brother to come in as his first sergeant.

Bill took the post. It meant a promotion, and it kept him in the regiment. He'd have more money to send home, and he'd still be around most of the action.

I'm not sure whether he valued the promotion or simply a new challenge. I know he thought highly of Dawson—a wise assessment, as it would turn out. G Company had suffered enormous casualties; leaving Sicily, it was listed at 60 percent strength. Rebuilding it meant it could take on the personality of its leader, an aggressive, hard-charging GI. But it also meant a lot of training. Bill and the other NCOs had their work cut out for them.

Around this time, Captain Dawson sat on a court-martial of a GI who had deserted just before the invasion of Sicily. The court-martial found the man guilty and sentenced him to death. (The sentence was commuted to forty years.) Dawson's sentiments in a journal he kept spoke for many in the army:

> *One lives a lifetime in a matter of seconds on the battlefront and many are those who have paid with their lives in order to protect the precepts of life for which we fight. Is it then fair to them to let one enjoy these blessings that shirks his responsibility and duty*

when others have paid so dearly? My only regret is that our judgement was not upheld . . .

A lot of things can be forgiven in war; letting the guy next to you down isn't one of them.

SLAPTON SANDS

With the holidays gone, training started picking up. Around the middle of February we moved to Camp Devon, on the southwest coast. Two or three months earlier, the three thousand people living in the villages around Lyme Bay were ordered to leave. They weren't told why, but many probably guessed that the beaches were a good place to practice amphibious landings. The cliffs and waves were similar to those across the Channel.

When most Americans think of beaches, they picture soft sand. But the beaches on both sides of the English Channel often consist mostly of rocks—"pebbles" ranging from fingernail size to something hard to get your whole fist around. The British call them shingle beaches. Many beaches—including those in Normandy where the planners were thinking of landing us—were a mix of sand and shingle. Slapton Sands, with pebbles mostly the size of a schoolboy's marbles, was one of these.

It also had cliffs above the beach—exactly like Normandy. And its location was thought to be relatively safe from German interference.

We traveled there March 11, spending the next several days practicing full-scale assaults on the beach: rope ladders, landing craft, live fire. There were some twists—tanks with funny shapes and large funnels to help their engines breathe in the surf, for example. But most of what we were armed with and much of what we did was familiar.

Infantrymen were equipped with Bangalore torpedoes—long, slim pipes that contained explosives and were slipped in or under barbed wire or other obstructions, then exploded to clear a path. GIs carried BARs, or Browning Automatic Rifles, an early light machine gun, along with Thompsons to increase firepower.

We lugged packs and medical bags as we sprinted from the landing craft, ducking the mock gunfire like the infantry, then racing to treat the wounded. Obstructions, barbed wire, and simulated mortar and artillery shells were all a big part of the drill.

It was a refresher course for those of us who had landed in Africa and Sicily, a reminder of what it was like to hit a hostile beach. For the new guys, it was an introduction to rocking waves and overturned stomachs, the nauseating smell of diesel engines and the heartless whistle of mortar shells.

It barely hinted at what the real invasion would be like, but it was better than nothing.

Slapton Sands saw a series of large-scale rehearsals in the weeks that followed. Then came tragedy.

At the end of April, German navy commanders picked up on heavy radio traffic in the area. Patrol boats sent to the area spotted eight LSTs—Landing Ship Tanks, immense craft that, as the name implies, were designed to take tanks and other vehicles ashore. The German commanders launched an attack. Three of the LSTs were hit. Two were sunk, one submerging quickly while the other burst into a spectacular set of flames that lit the night.

A total of 749 men, including some 198 sailors, died in the attack. The toll might have been even worse had the Germans realized they were not fighting destroyers, as they originally thought, but unarmed landing craft. An earlier accident between escorts had left the boats with practically no protection, and they would have been no match for the speedy little killers.

For some reason, we don't often think about the fact that soldiers die in training. Rarely do we note their sacrifice. In this case, it took more than forty years for the army to get around to formally honoring the men who had died that night. But their deaths were as much a part of the war as ours.

Medics were part infantry and part medical department, the point of intersection for both. On the ground with the infantry, we fit in at the company and battalion level, working directly with the men on the front line. At the same time, we were the forward part of a much larger team. And that team was working as hard as we were to get ready for the invasion.

Anticipating large casualties not only on D-Day but in the weeks that would follow in the battle for Normandy, the medical department ramped up the normal system for treating the wounded and getting them off the battlefield and into hospitals. The basic outline was the same we'd always followed, but there were many more facilities, and a few more steps along the way.

When someone was hurt, his company medic would treat him; if the injury was bad enough, stretcher bearers would take him to a battalion aid station very close to the forward line. There he would be evaluated and treated. As always, the battalion aid stations would be as close to the action as possible. There would be at least one doctor and often two who could treat the most serious wounds. The more severe cases would be stabilized for further transport.

Soldiers who were too wounded to return to their unit or who needed more treatment would go to a collecting station and then a clearing station in the rear connected to a field hospital. Though housed in tents rather than permanent buildings, the field hospital had enough advanced equipment to deal with the most serious wounds, which we anticipated would be sucking wounds to the chest threatening a man's ability to breathe, and stomach and abdominal wounds. There would be X-ray machines, sterilization—basically everything you would find in a first-rate hospital emergency room, and maybe a little more.

From the clearing station or the field hospital, the injured would be evacuated to England. Landing craft were being adapted to serve not only as transports for the wounded but as floating emergency rooms. LSTs could take upwards of two hundred passengers.

The LSTs and any other ships carrying the wounded would be routed to different ports on the English coast. Once they arrived, the wounded would go to a holding hospital close to the pier. There they would be evaluated again and treated. They would then be shipped farther north, to either a transit hospital or a general hospital. Further care, if needed, could take place in the States.

Starting back in France with the first medic who treated him, notes were kept on tags and attached to a man's shirt to keep track of his diagnosis and treatment. These records would stay with him all the way to the hospital.

Some five thousand ambulances had been mustered for this network in England; special trains were prepared and put on standby. A large number of African-American troops who had been assigned to service companies as drivers, stevedores, and the like were put on as stretcher bearers for the transports.

Like much of the rest of America, the U.S. Army was segregated at the time. With very few exceptions, African-American troops had not yet been allowed into direct combat roles. They would eventually prove that this great injustice was a major mistake, but that lay in the future.

WORLD AT WAR, SPRING 1944

Those of us in the 1st Division were focused pretty intensely on France, since we knew that would be our next destination. But this was a world war, and there was more to it than us.

Every night, Royal Air Force bombers left from bases in southern England to strike strategic targets in Germany. American planes took off in the day to do the same. The Battle of the Atlantic—the fight between U-boats and convoys—continued, despite the large losses the Germans had suffered in May.

In Italy, Mussolini had resigned and been arrested, only to be liberated and returned to power as Hitler's puppet. The rocky hills and mountainous terrain of central Italy was a force multiplier for the German troops who had rushed in to fight the Allies. They made the Fifth Army under General Mark W. Clark struggle mightily for every inch of ground. The Allies had Rome in their sights that spring, though there were no illusions that taking the capital would end the war there.

In the Pacific, the U.S. Navy had won a major battle in June 1942 at Midway, sinking four aircraft carriers. That victory marked the turning point of the war, which was slowly becoming clear. By the end of 1943, U.S. forces had taken Tarawa and Makin, rolling back Japanese gains. The spring and summer would see us invade Hollandia, Saipan, and eventually Guam. MacArthur was planning his return to the Philippines.

In Eastern Europe, the Soviet army had begun to make serious gains with counterattacks following the German defeat at Stalingrad at the beginning of the year. Leningrad, which had been surrounded but not taken by the Germans, was relieved in January; the summer would see two massive offensives practically destroy the German Center and South Armies.

Out in the deserts of the American West, a team of scientists were working on a weapon that would radically change warfare and geopolitics. We knew nothing of that, and to be honest, didn't know much about the rest of the war aside from what we might glean from *Stars and Stripes,* the army's GI newspaper, and BBC broadcasts. New equipment kept arriving, and more men were assigned to us. We didn't need news stories to tell us rest time was nearly over.

By the middle of May, unnecessary travel by civilians to the south of England had been banned. We got new equipment—six sandbags for each vehicle, to be used to bulk up protection from mines.

General Huebner addressed the division on May 19, 1944, telling us in broad terms that the invasion would kick off soon, and we were to be part of the spearhead. The following week, the officers were briefed on the plans.

The medics were issued a second med kit; we'd have two going into the beaches.

We were also issued gas masks, and told to practice wearing them.

We were supposed to wear them for an entire day. I settled for twenty minutes.

I don't think I even took mine with me when we went to Normandy. I know none of the guys wore them that day. They were the most uncomfortable and impractical things you could imagine. Just as bad were the heavy protective suits we were issued. Get one of those wet and you'd never make it through the surf.

Those also didn't get worn.

Other equipment made more sense and was put to use. We waterproofed our Jeeps, upgrading the wires and filling openings with putty. We were issued sandbags to put along the floors, providing a little protection against mines.

On May 17, we set out from Bridport to a way station about halfway to Weymouth Harbor. There we were quarantined—no one could leave, and no one who wasn't part of the unit could enter without special permission.

I saw estimates on how many casualties we'd have to deal with. The numbers were high; 15 percent was the average. Some planners were talking even higher. We were clearly facing the most difficult battle we'd ever been in. There would be a lot of guys we'd have to save. And many we wouldn't be able to.

The plan for us was to board a troopship, which would anchor roughly ten miles from the Normandy coast. At the appointed hour, we would load into Higgins boats—thirty-man landing craft with swing-down ramps at the bow—then hit the beaches. Our regiment was part of the first wave, and I'd be on one of the first craft, together with a handful of guys and a doctor to set up a first aid station. It wasn't going to be a full-on aid station; we weren't so optimistic to think that we'd get a tent up. We just planned on finding a safe place to treat guys out of the water. A bigger aid station would be set up a few hours later when Major Charles Tegtmeyer, the regimental surgeon, came ashore with more medics.

Our ship, the USS *Henrico,* was a simple merchant vessel converted to troop carrier duties. It had a single stack midship, with a bare-bones superstructure and no-nonsense furnishings; originally a cargo carrier, it was all business and had been from the day it was commissioned, unlike our previous transport.

The entire 2nd Battalion went aboard the ship on June 4, expecting to invade the next day. Once settled in, we had detailed briefings using sand tables that showed the defenses and gave us an idea of the geography of the beach. We tentatively planned where we might site the aid station.

We also saw that the 16th Infantry was going to face some of the roughest defenses on the beach. We might have expected that, given our experience.

It was a mixed honor.

The weather was so bad that Eisenhower delayed the landing twenty-four hours. Some of the other ships had already left the dock when they had to be called back. That didn't happen to us, but it was small consolation. We sat and waited through the night and most of June 5. The weather just a few days before had been perfect spring, with a clear sky and a warm, light breeze. That morning, it was vicious.

We didn't know it, but the poor weather had nearly closed the window for our invasion. If the forecast didn't improve, we'd have to sit around for another few weeks before trying again. In fact, the delay might change plans entirely—the element of surprise was at least part of what Eisenhower was banking on to make the invasion a success.

The weather didn't look much better that evening when the ship finally pulled out of the harbor. Fortunately, there were other things to think about besides the pelting rain and wind. A few guys played poker; you could always find a game before a battle. Other men looked at the Bible. Some prayed. A few cried.

LAST WORDS

We anchored about ten miles from the shore. The wind screeched above decks as I went out to catch some air and see if I could find my brother. It was just getting light—not dawn, but the false dawn that teases you with the promise of day.

It was the faintest tease—the sky was still overcast, and the air was so heavy I couldn't tell if it was spit from the waves or the beginning of a squall.

My brother saw me and called me over.

"This will be a rough one," I said. We agreed it would be far harder than Africa and even Sicily.

"How lucky can we be?" one of us said.

The other would have nodded, knowing exactly what he meant—both of us had landed twice before without a scratch. People had died all around us. If that was the sort of thing that involved luck—and our experience said it had to be—then there must be limits to that luck. You could toss a coin only so many times before tails finally came up.

We talked about Mom and Dad, and what they would think if they knew we were here.

"They'd worry about us."

"Probably, they're worrying now."

There was a call from below, and movement on the crowded deck. Word spread instantly, but we didn't need to hear what was being said, because we knew the meaning instantly.

Get to your stations! It's time.

Time.

I shook my brother's hand. That was us—no hugs, no fancy words, just hand on hand.

"If I'm lost, take care of my family," he told me.

"Same for me."

He nodded. I took a breath, then turned for my side of the boat, preparing to test my luck in the biggest battle of the century.

EIGHT

A Mighty Endeavor

RIDING IN

My guys were all coming up the stairs to the deck. I'd already strapped on my two med kits and the life belt; I walked with them to our station near the rail, waiting for our turn to climb down into the Higgins boat. The company guys split off, going to their units; each would ride in with the men he was to look after.

We had two nets to load the boats. Guys would go over the side one at a time and climb down, their descent steadied by a pair of sailors at the bottom who held on to the net, trying to keep the landing craft tight to the ship. With the waves, this was a near herculean task; they strained against the momentum not just of the sea but of the men jumping in.

It was so dark when I started over that I couldn't see the Higgins boat below. The ropes were thick and very wet, and even with my experience and practice I had trouble going down.

Higgins boats were about thirty-six feet long and eleven wide. They had a shallow draft, meaning that even when fully loaded they could get very close to shore. The front, or bow as the navy called it, needed about two feet of water; the stern, another foot.

In other words, if the boat beached, you generally got off in the water.

We could cram more than thirty guys in a boat; the craft were big enough to fit a Jeep, in which case you could get maybe a dozen men. Depending on the engine and configuration, they had a top speed of nine to twelve knots, or about ten to fifteen miles an hour to an army guy like me.

Each 2nd Battalion boat had between twenty-seven and thirty-one men, with at least one medic per boat. Our boat happened to have four medics, including a company aid man. Meyers was one, I believe, and Lepore; I can't recall if the third was Joe Baliga, or maybe Willard Braddock, or even Appleby. All were there on the beach with me, so if they weren't in my boat, they were in another somewhere alongside.

Dr. Fred Hall was with us, too.

The idea was that once we had a foothold on land, the three of us medics would set up an aid station. Not the full setup with a tent—there was little chance we'd have that sort of luxury in the first hour or two on the beach. But we would find a place of relative safety where we could take and triage the wounded, as we had in Sicily.

Two med bags, our personal gear, the life preservers, rain gear—the medics were as loaded down as everyone else. I don't remember any of my guys wearing the Red Cross symbols on our helmets or arms, but others did. By this point in the war, I'm sure no one believed they would offer any protection, or do anything but make us prime targets.

Seeing there was aid nearby in case you got hit meant some comfort, faint as it might be.

The last guy came down, jostling in to find a spot. Our boat moved off a short distance, joining two or three others that had already been filled.

Rain hoods flapped in the stiff wind while we waited for the others. Wet hands gripped the wet bags covering their rifle barrels. Men shifted as best they could, trying to ease the load of the heavy packs and other gear hanging across their backs and chests.

Guys cupped hands against the wind to have a last smoke. Others said prayers.

I was anxious to get going; I'm sure we all were. But at the same time there was a certain amount of dread, knowing what would come next. Unless we were very lucky, some percentage of the fellows crowded around me were going to die.

Will I bandage that fellow? Will I have to tourniquet that man's leg?

Any medic who stopped to think that way would have had a breakdown. You focus on your job; you push yourself back and do your duty, distancing yourself as much as possible from the human side. You wall yourself off.

If you don't, you're useless.

Yet underneath it, humanity stirs. You can't help it.

Between the waves, the diesel fumes, and what we were about to do, every guy on the boat got seasick, including me. I can't remember what we had for breakfast, but somehow it stayed in my stomach. I didn't throw up, but you wouldn't have known it from looking at my uniform; it was covered with vomit spray, thanks to the wind and the crowding in the boat.

Twenty minutes, twenty-five—finally, our flotilla was ready. The landing craft passed some sort of signal. Engines revved, the boats—and our stomachs—heaved as the sailors cranked their tillers, pointed themselves at the shore, and hit the gas.

We were too far out to see the land; as far as the soldiers standing in the well were concerned, we were heading to a compass point in the distance, hoping like hell the sailors knew their business. Waves crashed over the bow and sides, chopping at the boat as it raced forward, throttle wide open. The engine and the water passing beneath the hull set up a staccato vibration that drummed the soles of my feet.

Guys were still puking. With packs and guns and gear, we were crowded tightly enough that no man could fall over without working at it, but the waves shoved so hard that it was impossible not to jostle against your neighbor or feel that any second you would lose your balance and crash in a senseless, seasick heap.

The sun hadn't broken through the clouds, but the light was strong enough to see anything close in the water, and to at least make out the shadows moving with us toward land.

I was forward, at the left—port—side of the ramp, close enough so I could get out first. There was a slit of an opening at the center of the ramp near the top; in training, we'd peeked through it to see where we were heading. Now there wasn't much to see except the gray void before us. Behind me, guys craned their heads over the gunwale, stealing glimpses of what lay ahead.

I stood and waited.

OVERLORD

Our little landing boat was part of a vast armada destined for Normandy. Some three thousand landing craft were to ferry more than 150,000 men across the rocking waves to five separate sections of beach that day. The targets spanned fifty miles on the coast, from Merville on the east to Sainte-Mère-Église on the west. Paratroopers had landed a few hours ahead of us, seeking out strategic sites inland as they aimed to disrupt the German response and help us get off the beaches.

The British Second Army was landing at three areas on the east. The farthest beach on the east was code-named Sword and was northeast of Caen, an important day-one target. The British 3rd Infantry Division would be the primary force landing there.

The Canadian 3rd Division would be at Juno on the east side of Courseulles; next to them were the British 7th Armored and British 50th Infantry Divisions, which would strike at Gold Beach. Inland were the cities of Bayeux on the west and Caen on the east. Caen was strategic because holding it would open a route across northern France.

The two American beaches were U.S. First Army territory.

Landing on the east coast of the Cotentin Peninsula, VII Corps and the 4th Infantry Division would take Utah Beach. After the beachhead was secured, troops would strike out for Cherbourg, a port city at the top of the Cotentin Peninsula. Cherbourg was seen as a key port for supplying the offensive that would follow.

We were in V Corps, which included the 29th Infantry Division as well as the Big Red One, heading for Omaha Beach. Strategically, we were protecting the others' flanks, connecting the sectors, and striking inland to expand the Allied areas. The two American sections were a good distance apart, as were Omaha and Gold; our early objectives were aimed at closing the gap as quickly as possible, to prevent the Germans from outflanking the forces and counterattacking at their weakest points.

Omaha was roughly five miles long, running from Sainte-Honorine-des-Pertes to Vierville-sur-Mer. Cliffs rose on either end; the entire beachhead was crescent shaped, a rounded indentation of the shoreline. The beach itself rose gently from the water over a quarter mile or more to a narrow row of rocks or pebbles, sand dunes and seawalls; beyond this was a shelf of "shingle"—more small rocks—maybe two hundred yards wide, and then bluffs that were in some cases well over one hundred feet high. The exact dimensions of the beach depended on the tide; at high tide, there was precious little sand in front of the high ground.

No seaborne landing is an easy thing, and D-Day was no pleasure cruise for anyone. But our assignment was unarguably the most difficult, because of both the geography and the German forces that were known to be in the area.

It would turn out that there were far more defenders than army intelligence realized. Even the weather was worse than anyone had imagined.

The planners divided Omaha into ten sections, with every few hundred yards getting its own name, from Charlie to Fox Green. Our section was the widest—Easy Red.

I should say that "Easy" would have been chosen because of the alphabet, not as a remark about the defenses, ironical or otherwise.

Two battalions of 16th Infantry were landing in the first wave

Alabama, 1920s: Yours truly at left, feeling a bit shy that day. My younger brother Harland is on the far right, with family friends Billy and Lois Reeves when we were elementary school age. *(Courtesy of the author)*

My brother Bill *(left)* and I during the war years. We would fight together from the shores of North Africa to Omaha Beach and back. *(Courtesy of the author)*

"Beer party—16th med[ical] det[achment]": We trained hard, but we also had our moments of fun, such as this gathering in Massachusetts. Our war was still a few months away. *(Courtesy of the author)*

Here I am at the end of the war, Staff Sergeant Lambert, in uniform. *(Courtesy of the author)*

North Africa, 1943: Any vehicle a medic found could become an ambulance.

Often the wounds we treated in Africa were quite horrific. These aren't my guys, but they did the same.

The pipe a German pilot gave me just before he died in an olive field in Tunisia, Africa. I've kept it in my possession ever since. *(Courtesy of the author)*

Plasma was a lifesaver during the war, but combat medics rarely had the opportunity to administer it. Here Private Roy W. Humphrey is treated at a regimental aid station in Sant'Agata di Militello, Sicily, on August 9, 1943.

Some of the wounded being evacuated after the Husky landings.

General Terry Allen *(left)*, our beloved commander during the North Africa and Sicily campaigns, confers with General Omar Bradley.

"The Big Red One": the patch of the famed 1st Infantry Division.

General Huebner took over the 1st Infantry Division from Terry Allen in August 1943.

C O N F I D E N T I A L

HEADQUARTERS 1ST U.S. INFANTRY DIVISION
A.P.O. No. 1, U.S. Army

25 July 1943.

SUBJECT: Award of Silver Star.

THROUGH: Commanding Officer, 16th Infantry.

TO : Staff Sergeant Arnold R. Lambert, Medical Detachment, 16th Infantry.

1. Under the provisions of AR 600-45, as amended, Staff Sergeant Arnold R. Lambert, 7006617, Medical Detachment, 16th Infantry, is awarded the Silver Star for gallantry in action.

2. The citation is as follows: For gallantry in action in the vicinity of Mateur, Tunisia, 28 April 1943. Although subjected to heavy enemy artillery, mortar and machine-gun fire, Sergeant Lambert evacuated many casualties from a forward battalion aid station to an area where further medical treatment could be rendered. Residence at enlistment: Selma, Ala.

By command of Major General ALLEN:

Leonidas Gavalas

LEONIDAS GAVALAS
Lt. Col., A.G.D.
Adjutant General

C O N F I D E N T I A L

The citation for my first Silver Star. It's the only official Silver medal document I have, and thus the only one I list. *(Courtesy of the author)*

One of the most famous pre–Normandy invasion pictures: Eisenhower giving paratroopers a pep talk. *(U.S. Army)*

The D-Day invasion force crossing the English Channel. *(U.S. Army)*

Another famous photo of the beaches, with a good view of a landing craft such as the one I was on. *(U.S. Army)*

Landing craft approaching the Normandy beaches on D-Day. The first waves have already engaged. *(U.S. Army)*

I believe this image is of men from the medical battalion that came into Omaha after I was wounded.

These are not my men, but the image gives you an idea of the conditions on the beach even after our guys managed to move inland. *(U.S. Army)*

The Philadelphia Inquirer
PUBLIC LEDGER
An Independent Newspaper for All the People

WAR EXTRA

INVASION

Allies Land in France, Smash Ahead; Fleet, Planes, Chutists Battling Nazis

SUPREME HEADQUARTERS, ALLIED EXPEDITIONARY FORCE, June 6 (Tuesday) (A. P.). —Allied forces landed in northern France early today in history's greatest overseas operation, designed to destroy the power of Adolf Hitler's Germany and wrest enslaved Europe from the Nazis. The German radio said the landings were made from Le Havre to Cherbourg, along the north coast of Normandy and the south side of the bay of the Seine.

Allied Headquarters did not specify the locations, but left no doubt whatever that the landings were on a gigantic scale.

'FULL VICTORY' ASKED

Ringing in their ears, the American, British and Canadian forces who made the landings had these words from their supreme commander, General Dwight D. Eisenhower:

FLASHES

FOUR PARACHUTE DIVISIONS LAND

NAVY, COAST GUARD, MARINES IN BATTLE

DIEPPE HIT BY AIR, NAZIS SAYS

ALLIES GAIN 5 MILES IN ITALY

BERLIN CLAIMS TWO SINKINGS

LEAVE HOMES, DUTCH TOLD

26 WARSHIPS REPORTED OFF COAST

GENERAL DE GAULLE IN ENGLAND

WHERE ALLIES HAVE LAUNCHED INVASION AGAINST FORTRESS EUROPE

Landings Unopposed, Eyewitness Asserts

Big Allied Move Began May 28

Eisenhower Summons French People to Posts

Allies Order Coast Evacuated

Roosevelt Asleep When News Broke

Norwegian King Warns on Revolt

AIDED BY PLANES, WARSHIPS

TAKE ADVANTAGE OF TIDES

1st Communique

The newspapers reported the landings in the biggest possible type. Because of censorship and other factors, the early reports were far from 100 percent accurate.

Somewhere in Belgium
Dec 14th 1944.

Dear Donald

Just received your letter of
Nov 1st last night — it has
just taken a week to reach
me —
Decorations are terribly mixed
up — The Division is rear the
behind — You have a Silver Star
coming — Colonel Seitz put it
in — but it was lost so speak
to me about it then Ann &
I'll take care of it — I'll
also — do something about a
Bronze Star — & if & when
there is a Combat Medics Badge

"You have a Silver Star coming": a letter I received from Major Charles E. Tegtmeyer, regimental surgeon in the 1st Division. This would have been my third Silver Star. *(Courtesy of the author)*

A small portion of the American cemetery above Omaha Beach. Never forget.

A few of my medals and insignia, on display in my living room in North Carolina. *(Peter Hubbard)*

My wife, Barbara, and me at home, 2018. *(Courtesy of the author)*

The French town of Colleville-sur-Mer very kindly honored my fellow medics and me with a ceremony in October 2018. Below, D-Day survivor Charles Shay and I meet with two local children on Omaha Beach. *(Courtesy of the author)*

"Ray's Rock" on Omaha Beach. This rock saved many lives. *(Courtesy of the author)*

The plaque honoring our medics, now permanently attached to "Ray's Rock." Every man a hero. *(Courtesy of the author)*

at Easy Red and its adjacent sector, Fox Green, which marked Omaha's eastern boundary, on our left going in. The beach was below Colleville-sur-Mer; there was a draw or exit from the beach that ran to a road connecting with Colleville and other nearby towns; a second, narrower trail did as well. We wanted to grab these exits right away; take them, and a path would be open for us to get off the beach.

H-Hour—the targeted landing time—varied, depending on the beach because of the tides and other factors. Ours was 0600—6:00 A.M., roughly an hour ahead of the planned landings by the British and Canadians. Just ahead of us, specially trained engineers were to blow passages through the maze of obstacles and mines that littered the approach. By then, heavy bombardment from airplanes and ships in the channel would have neutralized the big guns and bunkers on the bluffs and beyond.

We'd be off the beach and driving our Jeeps and trucks through Colleville two hours after we landed.

At least, that was the plan.

Often we talk about what happened on D-Day as isolated events, but the entire assault was a complicated, interconnected event. We were depending on the other companies and battalions involved just as they were depending on us.

And there were many other components besides those from the Big Red One. Nearby to our west, the 2nd and 5th Ranger Battalion was to scale the cliffs at Pointe du Hoc. Intelligence had spotted batteries of 155-millimeter guns and some smaller weapons at the top of a nearly sheer cliff; the massive shells from those guns could wipe out an entire platoon with a direct hit. The Ranger force—only a few hundred men, all told—were to land below du Hoc a few minutes before we got to the beach, climb the cliffs with rope ladders, defeat the infantry guards, and spike the big guns.

Looks impossible, even on paper.

At least it could be said that the army had come to expect the Rangers to do the impossible, thanks to their track record. I'm not sure how anyone could have expected tanks to float.

But that was the idea behind the DD tanks—specially prepared Sherman M4s that were to accompany us to shore.

Not land in shallow water. Actually swim.

The tanks had specially sealed, hull-shaped underbellies and propellers. To the naked eye, their most telling features were the "flotation screens"—a series of tubes around their middles that looked for all the world like a child's swimming ring.

The tanks were seen as one answer to the problem of German firepower onshore. Command believed that they would help neutralize German strongpoints, providing firepower as the infantry swarmed through cracks in the defenses.

It was a great theory, if they could actually swim. Thirty-two DDs were scheduled to join us in the first wave at Omaha; another two dozen conventional Shermans were to come in the following waves. (DD stood for "Duplex Drive," a code name for the project.)

You'll have to excuse a medic for thinking that a thirty-four-ton tank has no business swimming.

Still, in my opinion, the idea wasn't nearly as out there as the decision to put parts of the 29th Infantry in the first wave.

I mean absolutely no disrespect for my brother soldiers. The

116th Infantry Regiment, tasked for the job, had good men. But unlike the regiments in the 1st Division, this was their first battle. And the 29th was a National Guard unit; the bulk of its members had not been full-time soldiers before being mobilized.

That is not a criticism of them. It's simply a fact—their unit was not designed to be on the leading edge of a desperate assault. A lot of effort had been put into improving divisions of the National Guard at the start of the war, and the 29th's preparation was certainly far better than the norm a few years before. But its inexperience made it a very poor choice for the first wave.

The first time in combat for any unit—including ours—was always difficult. Commanders didn't know how their soldiers would react. Privates didn't know which NCOs to trust; NCOs would be unsure of their officers. A thousand things you couldn't learn from instructors or on maneuvers would become important in the flash of a gun.

And there was that killing thing. It wasn't an easy thing to learn, even if someone was shooting at you.

True, they were between the Rangers and us. But surely a more experienced unit could have been found. Our other regiment, or part of the 4th, or another unit that had tasted action in the Mediterranean.

From what I understand, the decision was political, made because the head of the division wanted to prove that National Guardsmen were the equal of "regular" army.

They certainly could be, but only after they went through what we had.

Which would soon happen, in spades.

WHAT ABOUT US?

The first shell passed without us noticing. The second and third must have as well. Short or long, left or right—our senses were dulled by the rumbling tumult of the Higgins boat charging toward the beach, and it took long moments before we realized we were being fired at.

The flashes in the sky suddenly made sense. This was the moment when your stomach either tightened or let loose altogether. This was the moment you steeled yourself and decided you were doing your best. Either that or everything let go, your legs became water, and you lost your will to do anything more than breathe.

Either way, you were getting the hell off the boat. We all were.

Behind us, our battleships were launching salvo after salvo. A destroyer passed nearby—I saw the shadow—and its guns flashed.

I thought how incredibly close he was to the shore. Surely he would be hit.

If he could be hit, what about us?

Moving at top speed, the landing craft twirled up whitecaps along their sides. Specially modified versions sent volleys of rockets skyward, rows at a time; the engines all seemed to ignite a few feet above the ship, a burst of explosive energy as coordinated as a symphony.

Our boat veered hard right, apparently steering away from an obstruction one of the sailors had spotted at the last moment. The Germans had seeded the long beach approach with girders and mines intended to stop us before reaching the water's edge.

The engineers had been tasked to sneak in just ahead of us and clear out a path; they'd missed that one.

And another; our boat veered back.

And another!

Where are the engineers? Where are our guys?

I didn't realize it, but the answer was rumbling around us, encoded in salvos from the German batteries on the bluff.

They're dead. You're on your own.

I raised my head and looked forward, trying to make out what was ahead, how far we were.

Before I could focus my eyes, something rattled against the ramp. I hunkered down automatically, ducking; my legs and chest, my neck and upper body were all quicker than my brain to realize what was happening: a machine gun had targeted our boat.

We're close now.

When the rattle stopped, I rose again, leaning to the side to look out the boat.

There was smoke everywhere. A thick fog of it wafted down the dunes ahead. I couldn't make out the waterline. There were flashes, dots of light everywhere. Enemy gunners wailed away.

The water ahead erupted with fury . . . mortar shells.

Smoke, and more smoke.

Diesel in your lungs. Retch in your mouth.

The uniform wet against my thighs.

Sshhwosh, sshhwosh—the massive shells of the battleships had been redirected, firing farther inland to avoid hitting us. The symphony had gone up an octave.

The landing craft to our left slowed abruptly, bow nudging down.

Hit.

Another, farther on, had stopped. Gray men rushed out, ran, fell, danced in the surf.

"When the ramp goes down, run and try and get under the water," I told the others. I yelled, trying to make myself heard over the roar of the engines, surf, and explosions. "Get into the water where the machine gunners won't see you!"

Did they hear me?

I'm not sure why I spoke. They knew what to do as well as I did.

Did I even speak? I can't remember now, not for certain. What I do remember, most vividly, deep in my bones, is this:

The boat lurched forward. The ramp went down, and I launched myself forward.

AROUND US

We were supposed to hit the beach around six o'clock. I'd guess that by now it was six thirty, or closer to a quarter to seven. The entire wave was running well behind schedule.

At the far western end—on my right, as I looked at it from the water—the Rangers were struggling to get to their landing points beneath the gun emplacements at Pointe du Hoc. It would take another thirty minutes for them to finally make land; when they did, Lieutenant Colonel James E. Rudder would find he had a much smaller contingent than originally planned. Two destroyers offshore, the USN *Satterlee* and the British ship *Talybont,* were sailing well within the enemy's range, attempting to neutralize the defenses with their cannons.

Another group of Rangers and companies from the 116th Regimental Combat Team under the 29th Infantry Division were having a hard time finding their beaches. Some of the boats ran into sandbars well offshore. As the landing craft dropped their ramps, men found themselves wading in relatively shallow water; a few steps later, they were nearly drowning in water to their necks and above. Our companies in the 16th, my brother's among them, had similar problems.

The heavy weather had made it difficult for the bombers to see their targets properly, and the bombs intended for the heavy artillery behind the beach had missed. With no spotters ashore yet, many of the ships supporting us were having similar difficulties finding their targets.

Not only were most of the German defenders still in place, but so were nearly all of the mines and obstructions scattered in the low waves and beach. The engineers assigned to blow up passages had not managed to do so. Buffeted by the same problems we had, many were lying dead in the surf.

Farther inland, paratroopers from the 82nd and 101st Airborne Divisions had landed during the night. As had happened in Sicily, the winds and weather made for a very scattered landing, and many of the units spent hours finding each other and then organizing attacks.

The major exception came at Sainte-Mère-Église, where the 505th Parachute Infantry Regiment's 3rd Battalion took the strategic town a few hours before we launched. It was the first French municipality to be liberated by Americans.

Sainte-Mère-Église was a key point inland from Utah Beach on the Cotentin Peninsula, where troops from the 4th Division were heading. They, too, had trouble with the waves and wind. And as the flotilla of landing craft approached the beach, three of the boats that were acting as guides hit mines, in effect blinding a good part of the fleet.

The British were preparing to invade at Sword, the easternmost beach of the D-Day plans. Inland, a British parachute unit had already seized and destroyed five bridges that could have been used by German reinforcements.

Offshore at Juno Beach, the Canadians were boarding landing craft for the assault as destroyers made a last-minute run at the defenses. Here, too, weather caused delays, pushing back attack times and complicating plans to land tanks meant to support their attacks. It would be closer to 8 A.M. before the first wave touched land.

Gold Beach, the middle of the five beaches and the one to our east, was being lashed by British shells. The Tommies were scheduled to land at 0725, not quite an hour away. Like everyone else, they would have a tough time with the weather, and find that the beach defenses had not been damaged anywhere near as badly as they'd been promised.

The German command was still not sure what was going on. While the paratrooper attacks and the heavy bombardment had raised alarms, the German commanders away from the beaches were hampered by everything from the French Resistance to the conviction that we'd never land at Normandy, much less choose a day with bad weather and borderline tides. The men firing at the approaching landing craft had no doubts that they were under attack, but many in the upper echelons were unsure whether this was a diversion or the main assault. General Rommel—our old Africa nemesis, now in charge of the defenses in northern France—was only now being informed of the attack; he was many miles away in Germany, having gone to celebrate his wife's birthday.

He, too, had thought the weather was so bad that no one in their right mind would try an assault today.

IN THE WATER

From the glimpses I saw as my Higgins boat plowed toward the Normandy shore, I realized that the obstructions on the beach remained. Not only that, but there were geysers everywhere around us and ahead. The guns and enemy we were promised would be destroyed were very much alive.

The landing craft were under heavy fire. Not just the others, but us, too.

The front ramp rattled.

Bullets?

The noise of war—the explosions, the waves, the diesels—crescendoed.

Our boat stopped abruptly.

Go!

Go!

As the ramp dropped ahead of me I pushed over it into the water, jumping off the side. I expected the waves to hit my knees but felt them hard on my chest, pounding even my shoulders, trying to push me back.

Something hit my right arm—a shell fragment probably, or a bullet smaller than a machine gun's. It went through my elbow, making a clean tunnel where once there had been skin and muscle.

Doesn't hurt. Don't stop.

I ducked and began moving awkwardly through the high water, the mud and sand below sucking at my feet.

I need to get to dry land.

Move! Faster!

The hardest thing about combat is the noise.

War sounds like nothing you're used to in civilian life. The landing craft's engines had shielded some of the shrieks and the awful explosions. Now I heard them fully, and felt the reverberations in my spine.

Bullets and shells rained across the deep surf. The water percolated, as if the earth were furious with us—not just us, but all of mankind.

The noise of war does more than deafen you. It's worse than shock, more physical than something thumping against your chest. It pounds your bones, rumbling through your organs, counter-beating your heart. Your skull vibrates. You feel the noise as if it's inside you, a demonic parasite pushing at every inch of skin to get out.

I kept moving. Stopping was not an option. It wasn't a case of being brave or even thinking about what my job was. Stopping was dying. There wasn't free will or even a possibility of making a decision; I just moved to the beach. That's what I was doing.

Stopping was dying.

As I moved forward, I began focusing on a machine gun up on the bluff ahead, a little to my left. It became almost a landmark, its flashes like a beacon.

A deadly one.

Curved slightly. Not aimed at me.

Stopping was dying.

There was another machine gun on the right, some distance away. Between them, they had our sector of the beach well covered. They ground away, spitting lead at the men swarming from the boats.

There must have been other gunners on that hill, along with spotters and snipers, but the machine guns are what stand out now as I look back. The awful size of their bullets; the quick tap as the gunner touched the trigger; the holes they made in the bodies of the men I'd known and taken care of for years.

I hit a deep stretch and slipped under the water, hoping that would make it harder for anyone targeting me. The bottom rose abruptly; no longer ducking, I avoided a pair of obstructions, kept my balance somehow, pumped my legs to get more speed, any speed.

I still had that machine gun on my left. There were landing craft behind me, to both sides, men running and crawling.

A body floated nearby. The first I'd come across, or at least the first I realized was a body.

I stopped to help, but the soldier was clearly dead, facedown, torso ripped apart like a paper doll.

There were live guys moving, others floating. Those were the ones I had to care for.

I made it to the beach, dropped my pack, and got my wits about me, focusing on my job, not myself. A guy about fifty yards from me struggled in the water. His arms were up; he was yelling as a wave crashed over his shoulders.

Why didn't he just move?

I half ran, half swam to him. I grabbed and tugged, unsure why he was paralyzed.

He remained fixed there.

He's hung up in barbed wire!

There was a run of barbed wire along the bottom of the approach right where he was. His life preserver and pack had snagged in the barbs.

I grabbed hold of him again and pushed, trying to maneuver up as well as forward.

That didn't work. I held my breath and ducked beneath the waves, fumbling with a hung-up strap until I was out of breath.

Didn't work.

I took a gulp of air and ducked back down.

How can I free him?

I have to free him!

He wouldn't come loose. Finally, I went down again, and this time realized the solution—if the life preserver wouldn't come loose from the barbed wire, he had to come loose from the life preserver.

I pulled that and his pack off.

Gasping for air, I tugged him away. We moved toward the sand and pebbles at the shoreline, toward the raging gunfire.

There were dozens of men on the shore now, most ducking but a few firing.

The guy I'd pulled out had lost his rifle. I pointed him toward the shore as we got closer.

"Grab a gun—there's plenty," I told him. "Find your company or just join in."

He must have understood, though I can't say how he would have even heard my words in that roar. He pushed toward shore as I turned around to see who else I should help.

VISIONS OF HELL

We talk about Omaha Beach and think about it as a broad expanse of sand, but I spent much of my time that morning in water several feet deep. That was where a lot of the wounded were. The Germans had the area well mapped for their guns. The obstructions were mined. And the ground under the water was very uneven; the depth changed within steps. With your heavy gear, even if you weren't shot or hit by a shell, you could easily get in trouble well before touching dry sand.

The entire area was studded with obstructions; from a distance it looked like a young orchard in winter, each tree with its branches shorn off and a mine or two placed at the top. It wasn't safe to use them as cover, and not just because many were mined or relatively narrow. The Germans on the hills above had practiced using them as targets, sighting weapons on them.

The tide kept coming in, which meant there was less and less actual beach as the minutes passed. Any spit of dry land was deadly, nearly bare of protection, and well measured by the mortars and artillery behind the bluff, to say nothing of the machine guns and riflemen we could see.

A lot of the guys had so much equipment on that they couldn't stay upright once in the water. The life preservers were belts you put around your midsection. While they did supply buoyancy, they also threw you off balance in the water, especially if you weren't used to them.

The guys with the worst problems often had on two belts. Weighed down with all their extra gear, it probably seemed like a good idea to add an extra life preserver to counterbalance it. But what that did was tip them like a seesaw when they got in the water. Their upper bodies had all the weight; their bottom halves were lighter. The belts ended up helping to hold their heads under. Even one belt could be a problem if you lost your balance suddenly and became disoriented.

I saw a man like that and pushed over to him. Someone else came near and together we got him upright, got the life preserver off or cut loose, and pushed him toward the dry beach.

Bodies bobbed with the tide. A wounded man crawled up out of the surf, collapsed, turned over, and stared upward, a vacant look in his eyes. A group of four men ran from a boat toward the beach. One fell to the ground, shot. A moment later, a second one fell. The first stumbled to his feet and went on; the second never moved, not under his own power, not in this lifetime.

I went back into the water, helping a couple of other guys in the surf. I came across one who'd been wounded already and pulled him into shallow enough water where I could get an idea of how bad the wound was. He was lucky—relatively. It wasn't that bad. I got him onto land and went back.

Bandaging was tough with the water lapping at you. The soldier would be wet, the bandages would be wet, you'd be wet. I'd tuck the wrappings around the wounds as much as possible, wrapping an arm or leg or even a chest, and hope it stayed until we could get him more care.

Transfusions here were next to impossible. Things were just too wet and the gunfire too steady. The best hope in those first minutes was simply to save a person from drowning and bleeding out, get him some shelter, and then hope things improved quickly so we could give further care or get him evacuated.

Many of the guys who were wounded right off the boats stopped at the edge of the water, just on the dry land, too exhausted to move any farther. That made them easy targets. Worse, with the tide coming in, they were soon in danger of drowning or being swept away. I started moving among them, and came across a fellow with a badly damaged left arm. Every time a wave came in, his arm would swing up; he'd grab it with his right arm as it swept back.

I pulled him onto the dry sand.

The only thing holding that arm on was skin. I rolled out enough bandage to tie it to his chest. Then I dragged him up along the pebbles, gave him a shot of morphine, and went back into the water.

The next guy I helped was wounded too bad to fight, but he was able to move toward shore on his own power. I helped him, looking for someplace to send him that might be a little safer than the surf. I spotted a rock—really a stubble of concrete left from some sort of fortification—nearby. It was a good bit from the water, at least at that point, and while it was only a few feet wide and probably half that high, it was the only shelter I could see anywhere nearby. I pointed him in that direction, then left to help others.

By now, the second assault wave was coming in. It was a replay of what we had gone through. I pulled guys in, sending some of the walking wounded toward that rock—only to realize that the machine gun directly above it was zeroing in on that area, tearing apart anyone who wasn't fully protected.

This must be what hell looks like.

WN62

Ahead of me on the beach, Lieutenant John Spalding was leading E Company in the direction of the machine gun on the left I'd used as a marker. Their target was Widerstandsnest (Resistance Nest) 62, a strong point on the hill that not only had a good line of sight to the beach but also protected the key exit on that side.

Spalding and his men found their path upward blocked by heavy gunfire. They detoured, backtracking slightly and finding a way to cut through a nearby ravine that hugged the hillside, making it impossible for the gunners above to see or target them. Threading a path through a series of mines, they were stopped by fresh gunfire from above the ridge. One of the GIs got up and fired a bazooka shell at the nest; the rocket missed.

They reorganized. As covering fire from a Browning sprayed the position, the men charged upward.

They found a single man at the weapon, hands up, trying to surrender.

He turned out to be Polish, a prisoner of war impressed into the German army. Company E's first sergeant, Philip Streczyk, who'd grown up in a Polish neighborhood in East Brunswick, New Jersey, was able to translate enough of what he was saying to realize there were other Poles with the man above, and that they, too, wanted to surrender.

Dubiously, the company moved up. The man had been telling the truth. The area around WN62 was now in their control.

I don't know for an absolute fact that WN62 and the guns in that area included the machine gun peppering the rock on the beach where our guys were taking shelter. I do know that the firing there stopped. If Spalding, Streczyk, and the rest of E Company weren't responsible, then I'm just as thankful for whoever else did the job.

Taking out that one set of guns didn't make the beach a safe haven. We were still getting pounded by artillery and mortars, machine guns and rifles elsewhere on the bluffs. The other machine gun I'd spotted earlier was still peppering the right side of our sector. Men were still falling in the water, dying even as I gripped their shoulders to help pull them ashore.

I hadn't seen any of my other guys until finally I spotted Meyers with a clump of infantrymen from his company. They were firing up at the bluff, trying to move through the barbed wire and other obstructions.

There were more machine guns scattered above. They'd pause, maybe to reload or because their barrels were hot and needed to cool down. The gunners would pick up rifles and use them, joining whatever other men were there with them.

A lot of guys were shot in the head, I guess by sharpshooters manning the cliff. Corpses—men I'd have recognized less than twenty-four hours before—floated in the water or lay scattered on the beach. Some were the engineers who'd come to blow up the obstacles. Medics had gone with them, but been cut down before they could do their work. Not a single engineer that I saw had been treated for his wounds.

Guys would sink, weighed down by their gear. Legs floated up from the bottom.

I kept going in and out. More boats. More people near drowning.

My right arm had been bleeding from whatever hit me when I landed. It got so I could hardly use the arm, even though I couldn't feel pain.

Adrenaline.

Get the guys out of the water.

We hadn't planned on this, running in and out of the water, helping men manage the last hundred yards or whatever to land. You especially don't plan to treat someone in the water, or at least we didn't. We assumed we'd be on the dry land.

But whatever you plan doesn't matter. And there was no way I was going to set up an aid station on this section of the beach, not yet. The firepower was too great. The best I could do was that big rock.

Grab that guy. He's alive.

The pebbles on the beach would shatter when the shells hit. The splinters would bruise as they cut, as if you were hit by a baseball bat that punched a jagged hole at the center of the wound.

I ducked away from one of those pebble showers and saw a small area on the left or east side of the beachhead, a jumble of concrete and rocks—the ruins of a jetty maybe. There was shelter, at least more than the flat, open spots directly where we'd landed.

Five or six guys were already there. It was a better spot than my rock.

Send them there.

I'd been on the beach maybe twenty minutes; that's a guess, and not really an educated one.

A few minutes later, as I was heading toward someone with a red cross on his helmet, a shell hit nearby. I got slammed on the left side when it exploded. What hit me was a mystery, certainly then and even now. My uniform was ripped open; the flesh looked as if it had been pulverized and simply given way, exposing the tissue beneath.

Fragmentation of something. A helmet, I thought.

A boat blew up behind me. Maybe from that.

Left thigh, just above the knee, back of the leg.

Pain.

Pain!

I tore my pant leg enough to examine the wound. I worked up a tourniquet above it.

Tight.

Back in the water, I went to the nearest GI.

He was dead. I pushed him into the land and looked for someone else to help.

The cold water, I think, helped control or at least clean my wounds, but the pain was so great I had to reach into my med kit and take out one of the morphine-filled syringes.

I pushed the needle into my arm. I was beyond exhaustion. It was just a matter of time now before I gave out.

I kept going until I saw Ray Lepore, one of the medics who'd been on the landing craft with me, near the big rock. He was there with Meyers; they were going back and forth, really working, getting guys to shelter there, then working on them.

The spot on the left was better. I decided to tell them what to do, have them take over for me if—no, when—I became too weak. Which would be soon. Very soon.

You couldn't hear much except the explosions. You'd yell and gesture, communicating that way, but it wasn't anything close to normal conversation. It was caveman level, pointing and somehow knowing what the other person was trying to say.

Lepore grabbed a guy from the water who'd been hit in the chest. He could walk; we ushered him toward the big rock below the machine gun nest on the left.

"What can we do for him?" screamed Lepore.

Before I could come up with something, my medic and friend fell against me. His helmet spun to the ground, a foot away. A sniper's bullet had gone straight through, killing him instantly.

NINE

Beyond Despair

RESILIENCE

If despair was possible, it would have taken me then.

Why didn't it?

Ask that question of every man on that beach at that hour. Get a thousand different answers, each true for each, all true for all.

Duty. Courage. Fear.

Will.

Everything I had been through, from the Depression to my father's injury to my boyhood to my work experiences, to cutting trees to doing laundry, to joining the army, to Africa, to Sicily, to Tunisia, to Troina—not only had they brought me here, but they pushed me on, urging me to do more than I was capable of doing, able to withstand incredible pain and desperation.

To be who I was and what I was, a medic, a man assigned to help other men.

I got up and dragged Lepore's body six or eight feet up away from the water. Then I went back to work.

THE NATURE OF CHAOS

Gradually, the nature of the chaos surrounding me changed. It was still a mad jumble of noise, fear, and explosions, of boats coming in, mortar shells landing, of screams, exploding rocks, tears, courage, and sacrifice. But slowly these things clumped themselves into patterns. It was as if an invisible carpenter's chalk line had been snapped down for things to arrange themselves along. I began to understand the madness.

Men jumped from the boats in similar ways, again and again; they fell in different places but with the same wounds. Officers screamed the same words into the wind, then led charges that cut familiar paths.

The one constant was the noise, a mélange of screams, curses, and prayers that added timbre to the thundering explosions, the pop and snap of bullets, the whistle of passing shells. Boat engines, the pounding surf, even the wind—the sounds were amplified by adrenaline and heaving lungs. You wanted it to stop, but it was the one thing you knew must never stop, for if the beach grew quiet, it meant you'd been killed.

It's impossible for me to put an exact time on any of what happened on the beach. As near as I can calculate, this must have been past 7:30 A.M.; whether it was closer to 8 or 8:30 or even past 9, I have no idea.

By now, the regimental medical team headed by Dr. Tegtmeyer had come in. Realizing that our bit of the beach could not provide enough protection for a regimental aid station, they moved east, hoping for a place out of the worst of the gunfire. They eventually found a pillbox that had been knocked out; that became our first aid station, a few hundred yards from the rock and a little beyond the area I'd spotted earlier.

Those few hundred yards must have felt like a thousand miles to the walking wounded we sent that way, and even farther to the stretcher bearers. But it was the only thing to do.

By the third wave in, we were making progress, but it was difficult to tell from where I was, as the defenders were still killing our guys as they came in. G Company—my brother's—had a boat where every man on it was killed before they got out.

Every man.

Fortunately, my brother Bill was on another craft. But he, too, got hit shortly after landing, and by now was lying near the edge of the surf. Though we were maybe less than a hundred yards from each other as I scrambled and stumbled back and forth, I had no idea that he had been hit.

The leader of G Company, Captain Dawson, was climbing the ridge near the machine gun I'd used as a landmark on the right. Initially stuck by barbed wire in front of a ditch or tank trap, Dawson's team blew a hole through with a Bangalore torpedo, then worked their way up until Dawson, at one point, found himself stymied by another machine gun. He backtracked, went around, and got under the gunner.

Two grenades silenced the gun. The rest of the defenses there would soon be neutralized, opening an exit from the beach. But while the Germans had now lost key defenses on the hill, there were more scattered around, and plenty of artillery behind the bluff. Shells and gunfire continued to rain down on us.

At du Hoc, the Rangers had begun climbing the cliffs in front of the massive gun emplacements. When they reached the top, they engaged a small force there. In close fighting, they killed a number of men in the trenches, and got others to surrender after locking down the only exit from the bunker.

It turned out that the big guns weren't at the top of the cliff, or in a position to fire at all. The battery had been undergoing renovations. The Germans were changing the open gun emplacements to covered ones, and had taken the guns off their pedestals. The Rangers found them in a field nearby and quickly destroyed them. Then they returned to mop up a neighboring defensive site.

To the east of Pointe du Hoc on "Easy Green," another group of Rangers and the 116th Regiment were ashore after a confused landing that saw only one boat hit the right spot. Battered by gunfire and forced to wade or swim a hundred yards or more to shore, many of the men lost their weapons. Once on land, many found protection at a seawall, temporarily paralyzed by the heavy gunfire.

More landing craft came in, but the landings remained scattered throughout the morning. Slowly, and under constant fire, the 116th began to organize. An attack on one of the draws that led to an exit at the west end of the beach was too light to succeed; the troops regrouped.

Nearly all of the DD tanks that were to come in on Omaha had floundered. Most of our 105-millimeter howitzers, which could have provided considerable firepower ashore, hadn't made it either. With one exception, every gun of the 111th Field Battery sank offshore. The losses in the 7th Field Artillery Battalion and the 58th Armored Field Artillery Battalion weren't quite as bad, but we still lacked significant artillery on the beach.

Aboard the command ship *Augusta,* General Omar Bradley and his staff were becoming concerned. Reports from Omaha were sparse, a bad sign in itself. And as each minute passed, it became clearer and clearer that the landing's first objectives were not being met.

General Bradley would soon face a gut-wrenching decision: Should he give up on the landing, withdraw, and concentrate on Utah, where things seemed to be going considerably better? Or should he carry on with Omaha, doubling down with whatever reserves he could muster?

Either way, a lot more men were going to die. Each one would be on his conscience for the rest of his life.

Working my way across the beach, I found a man with a badly wounded arm. He was conscious, but bleeding so badly that I knew he was going to die.

I picked him up anyway, got him out of the water and over near the rock.

He looked at me with the question everyone asked.

"You're going to be okay," I lied.

I put him down and went on with my work.

THE RAMP

More boats were coming in; it must have been the fourth wave arriving by now.

I waded out, knowing what was going to happen, hoping things might be getting easier. A few yards away I saw a man out in the water; alive or dead, I couldn't tell. I went out to him and found him breathing. He was alive, wounded but not so seriously. But he looked too spent to swim.

The water might have been four feet deep. I got him with my left arm, hooking it around his.

While I'd been checking on him, a landing craft had raced over in our direction. Facing the beach, I didn't see it. Just as we started to move toward land, that landing craft dropped its ramp directly on us.

We went straight to the bottom, pummeled by hundreds of pounds of metal, then held there, pinned against the sand and rocks.

I'm going to die.

This is how I'm going to die.

I knew I was going to drown. I fought, but how do you fight some 26,000 pounds of steel, oak, and men?

The difference between life and death on that beach was slimmer than a hair. We'll never know how many guys survived because of some fluke of fate or twist in the wind. By the same token, we can't guess how many died for the same reason.

As I struggled below the water, pinned and hopeless, pushing against one of my own boats, a miracle happened:

The ramp went back up.

Free!

I suddenly shot to the surface, gulping for air.

The guy I'd come to save was still hooked on my arm. I leaned forward, hurting so bad that I sank to my hands and knees and crawled my way to the beach, dragging the GI with me. The fourth and fifth vertebrae of my back had been broken; I didn't know the details, but I sure knew the pain.

Meyers met me.

"I'm not going to be able to go any further!" I shouted. I told him to keep taking men to the rock, and to send the walking wounded to the aid station.

Then I passed out.

TEN

Deliverance

EVERY MAN A HERO

My fate then could have been the fate of so many others at that precise moment. Suffering from the loss of blood, back broken, leg torn open, and a hole in my arm, I could have easily slipped from unconsciousness to death.

Why I didn't is part mystery and part easy to explain. The easy half was my men—the medics around me saved me as I had saved others, checking my wounds, making sure I was stable, sheltering me from more calamity.

The mystery is everything else.

Especially this: why did the ramp on the LCVT pull back up and release me from the bottom?

Had it stayed down for only a few seconds more, I surely would have drowned, as would the man I'd come to save. That was a moment of fate. It passed in an instant, without time to be examined or questioned. It passed then, but remains present with me now, present always, an inscrutable fact I return to contemplate again and again.

Why did the ramp go back up?

Maybe the landing craft had come in at the wrong place on the beach. Dropping the ramp gave the men aboard a clear line of sight, they alerted the coxswain, and he backed out.

Maybe the ramp and boat swung with the waves or the wind, and that freed me. Maybe I owe my salvation to Nature rather than a man.

Maybe it was luck, good and bad mixed together. Maybe it has no meaning, just the interplay of chance. But I've come to believe God had a hand in it. For whatever reason, I was meant to survive that day. I was meant to do other things after storming the beach and helping my men.

I'm still working on what all those things may be.

I never found out what happened to the other fellow. But that's true of everyone else I helped on that beach that morning.

Surely, our paths have crossed. I've met them, and they've met me at the division and regiment reunions, at other events, maybe even in the supermarket. But the madness of those hours on the beach makes it impossible for us to recognize each other. I don't think you would even recognize your own grandma coming in, let alone recognize her years later as the person you helped.

There was so much gunfire, there were so many explosions, so many wounds. Men speak of being saved, but can't identify their savior.

And that's all right. In a way, it's better. Every medic who did his job that day was a savior; every man a hero.

GAINING CONTROL

I have no idea how long I lay unconscious. It was at least an hour, probably several hours.

A lot happened during that time.

First to me. Someone pulled me in behind the rock and made sure I was no longer bleeding too badly from my wounds. He wrapped my wounds, took care to protect me. Then he ran off to help someone else.

Meanwhile, small groups of GIs had managed to get off the beach and were slowly eliminating the enemy machine gun nests and other defenses on the bluff. But it was hard to tell from the beach. Smoke and confusion permeated the battlefield; you couldn't see enough of what was happening to tell exactly what was going on.

Back on the *Augusta,* where General Bradley and his staff were overseeing the operation, the first reports from Omaha were dire. Then things got even worse—no reports came back. As the morning went on, Bradley began to fear the worst. He had limited resources and reserves; throwing them into a hopeless battle could easily doom not just the overall invasion, but the conquest of Germany itself.

On the other hand, pulling out might doom the invasion as well. And it certainly would doom the men that would have to be left behind.

Desperate for an assessment he could trust, Bradley turned to one of his aides, Captain Chet Hansen. He told him to take a PT boat into the beach area and report back.

A bit like being told to drive over to Hell and see if Satan was still around.

Hansen did just that. The little boat ducked around exploding shells and got close enough to see that GIs were climbing through the bluff. Things might not be going according to plan, but they were going.

Hansen reported back. The battle at Omaha Beach continued.

The German force we faced at the water's edge was the 352nd Division. Intelligence had predicted we'd go up against only a regiment, not a division, and an overstretched one at that. With a few exceptions, the Germans were ably led and well trained. The crisscrossing defenses, the well-placed guns on the bluff, the coordination of the artillery and mortars behind the front line, all made Easy Red a killing zone.

1st Division persevered. After they took out strongpoints I'd used as landmarks in my mind in the first hour, the men with Captain Dawson and Lieutenant Spalding of the 2nd Battalion secured the hillside around them. Others followed. NCOs and lieutenants, squad leaders and platoon heads took the initiative to gather forces and lead them onward. In many instances, the men they led were not the ones ordinarily assigned to them. But they were making the most of the situation, acting out of native courage and training.

After a temporary halt in the landings as the beachhead got crowded around 0830, more reinforcements began arriving. Tanks now, and vehicles, as well as men and gear. The Germans in the middle of our front called for their own reinforcements; they were unavailable. Continuing to engage the Germans with everything from hand grenades to shells launched from battleships that weighed over eighty tons, we wore the Nazis down.

By 1130, the naval gunfire, fresh troops, and heavier weapons forced the Germans at the beach exit near Saint-Laurent-sur-Mer to surrender. To the left of the machine gun I'd marked on the eastern end of the sector when we came in, this was a clear path off the beach.

Once open, troops began pouring through. The momentum of the battle dramatically shifted; the 16th Infantry Regiment had taken the fight inland.

Omaha wasn't the only place Americans were making progress after some initial rounds of confusion.

Over on Utah Beach, our former assistant division commander General Ted Roosevelt was among the men who had been deposited as far as two thousand yards from their objective. This bit of bad luck had two bright spots:

One, the sizeable contingent of soldiers happened to land away from the main German force, which made for an easy landing.

Two, Roosevelt was not the sort of leader to stand around and fret over bad fortune.

At that moment, he was the only American general not only at Utah but on any French soil. He had not only volunteered but insisted on accompanying his troops, winning reluctant permission after making a nuisance of himself. And he hadn't done that to get stuck in the wrong place miles from their objectives.

Stomping around the beach in his trademark wool cap—the general disliked helmets—Roosevelt gathered the commanders who'd come in with him and plotted strategy. They could have gone back out to sea and gotten to the right place. Or . . .

"We'll fight from here," the general is supposed to have said. And they did.

Following his lead, the men cleared two paths off the beach, marched through flooded fields, and headed toward a key road onshore, achieving their objective with light resistance from the Germans.

Roosevelt's exploits that day were recognized with the Medal of Honor.

Meanwhile, other landings along Utah brought in more troops, these in the right spot, or at least closer to their targets. By the time I fell unconscious, the beach had been largely cleared of obstacles. Landing craft rolled in with almost no enemy fire to worry about.

The Utah landings had been a late addition to the D-Day plan, added by Eisenhower because of the strategic importance of the harbor at the top of the peninsula. While weeks of heavy fighting awaited the units that came ashore, on this day the landings were a spectacular success—not bloodless, but with lighter casualties than realistically expected.

Far to our east on Sword Beach, successful operations by the British 6th Airborne Division had put bridges over the Orne River and Caen Canal in Allied hands. The paratroopers, some of whom had landed by glider, managed to destroy the bridges inland at Dives River and neutralize the German battery at Merville. They paid horribly for their success at the battery, but this allowed the British landing forces to advance. Seven of the eight paths off the water area were quickly cleared; despite heavy fighting, British units went inland quickly and hooked up with the paratroopers.

Things would become slower and tougher the farther they

went. An attempt to hook up with the Canadians at Juno to their west stalled; later on, the British faced a serious counterattack from the 21st Panzer Division. The attack was the most ferocious coordinated counter by the Germans that day. The British managed to withstand it, eventually forcing the enemy tanks to withdraw.

On Juno, the Canadian 3rd Division fought through the German 716th Infantry; by ten o'clock Canadian reserves were on land, moving toward the train tracks connecting Caen and Bayeux. In the hours that followed, Canadian troops were within three miles of Caen.

DD tanks—landed conventionally in shallow water because of the rough seas—arrived in numbers at Gold Beach and played a significant role in the fight. But there, too, the weather and the usual fog of war, as well as the German defenders, played havoc with the assault in the very early going.

The tanks had trouble navigating in the muddy sand. And while some of the key big guns in the area had been hit by naval gunfire, the first wave found that much of the German defenses had largely survived the heavy bombardment.

Minefields were an obstacle—but also a chance to get through defenses. "Flail tanks"—Sherman tanks with what looked like large carpet sweepers attached to their front ends—cleared paths through the mines. A tank would move to the edge of the field, then start the carpet sweeper. Forty heavy chains attached to the roller out front would spin or flail, beating the ground and igniting the mines at a safe distance from the tank.

The tanks moved slowly, less than two miles an hour, and the chains had to be replaced after little more than a dozen mines were exploded. The vehicles kicked up enough dust to rival a sandstorm, and there was always a chance that a powerful mine might injure the crew. But the nine-foot-wide paths they cleared were safe for infantry to travel through. And the American-made Shermans had not given up their 75-millimeter cannons; these, too, were put to good use against the German strongpoints.

Fighting their way off the beaches with the help of the tanks, by mid-morning the British forces were well on their way to establishing a foothold roughly five miles deep in occupied France. On the right flank, the 47th Royal Commandos sprinted off the beach soon after landing and were heading in our direction, aiming to take Port-en-Bessin and hook up with Omaha. Unfortunately, they were to find the village well defended, stalling the hookup.

SO MANY HEROES

These brief descriptions, and even the vast libraries of books on D-Day, can't come close to describing what happened on the beach that day. There were so many acts of heroism, of men exposing themselves to enemy gunfire to advance, sacrificing themselves to help others—each one is a library of its own.

A dozen Medals of Honor were awarded for action that day; while that may be the largest number for a single engagement, it is a tiny fraction of the men who deserved one. The vast majority of heroes went unrewarded, at least officially.

I'll just cite the actions of one of my men, Stanley Appleby, as an example of what many did. Stanley, a T-4, was a polite, short, light-haired soldier, the sort who took any job I gave him without complaint. He never gave me a moment of trouble.

This is his citation:

The President of the United States takes pleasure in presenting the Distinguished Service Cross to Stanley P. Appleby, Technician Fourth Grade, U.S. Army, for extraordinary heroism in connection with military operations against an armed enemy while serving as a Medical Aidman with the 16th Infantry Regiment, 1st Infantry Division, in action against enemy forces on 6 June 1944, in France. Technician Fourth Grade Appleby disembarked from his craft some fifty yards from the beach under a hail of artillery shells and machine gun fire. A large number of casualties were sustained and, but for Technician Fourth Grade Appleby's prompt and courageous action, would have perished in the surf. With complete disregard for his own safety, he on numerous occasions waded into the surf to lead them ashore and immediately administered first aid. Despite the intense enemy fire, Technician Fourth Grade Appleby never slackened in his efforts to assist and render aid to the wounded on the beach. His intrepid actions, personal bravery and zealous devotion to duty exemplify the highest traditions of the military forces of the United States and reflect great credit upon himself, the 1st Infantry Division, and the United States Army.

It doesn't take anything away from Stanley to say that the actions of every other company aid man, every medic that I saw, could have been described with the same words.

Or with these, which describe George Bowen, a company aid man from Kentucky who was with one of our battalions:

The President of the United States takes pleasure in presenting the Distinguished Service Cross to George H. Bowen, Private First Class, U.S. Army, for extraordinary heroism in connection with military operations against an armed enemy while serving as a Medical Aidman with the 16th Infantry Regiment, 1st Infantry Division, in action against enemy forces on 6 June 1944, in France. As the men in the initial assault on the coast of France waded through the waist-deep water, a number were wounded and were in grave danger of drowning. Private First Class Bowen, disregarding his own safety, stopped in his efforts to reach the shore, waded through the mined and fire-swept water to go to a wounded man who was drowning and dragged the man to shore. He then proceeded to the fire-swept areas to administer to the numerous casualties. During the attack men were wounded in an assault on an enemy machine gun nest high on the slopes of a cliff. Private First Class Bowen, to reach these men, crossed an uncharted minefield and moved through vicious enemy fire to within fifteen yards of the enemy's machine gun nest to render first aid to the stricken men. Private First Class Bowen's intrepid actions, personal bravery and zealous devotion to duty exemplify the highest traditions of the military forces of the United States and reflect great credit upon himself, the 1st Infantry Division, and the United States Army.

Among the many lives medics saved that day was mine.

TAKEN OUT

At some point around midday I found myself floating in and out of a haze.

The sun shone on my face, bursting through the clouds and smoke.

I'm alive?

I barely knew. I opened my eyes and realized I was on the top deck of an LCT, one of the larger landing craft being used now as a seagoing stretcher bearer, taking casualties off the beaches.

It was warmer than it had been earlier, well into the 50s. The air smelled like diesel and the sea, spent explosive, and spent men. My back had been braced, but I could move my head enough to see around me.

I gradually regained enough of my wits to realize there were many men with me, lined up in the well where maybe an hour or so before, vehicles had been parked on their way in. Now we were heading back to England. The waves were still up, though nothing like they had been earlier.

A doctor bent over me. He must have asked some questions; maybe I answered. He bent down and looked at my dog tags and said something.

"Lambert? . . . We have another Lambert here."

The words cleared the fuzz from my head instantly. It had to be my brother; Bill and I were the only Lamberts in the 16th at the time, maybe the only ones in the whole division.

"How is he?" I asked.

There was the slightest pause before he spoke, the sort of delay that lets you know that what's coming next won't be easy to take.

"He's hurt real bad," said the doc. "We may have to amputate his arm and leg."

"Don't," I said. "Please. Don't amputate."

I was speaking for him, knowing he wouldn't want to live that way. If our positions had been reversed, he would have done the same.

The doctor didn't reply. I tried to move, grab him, do anything. I wanted my brother to be alive, but also to be whole, intact.

"Please," I begged. "Please."

The energy I'd mustered slipped out with the words. I sank back on the deck, exhausted and hurt, somehow feeling worse than I had those last moments on the beach when I'd lost consciousness.

ELEVEN

Breakout

BILL

They took me off the beach at Weymouth, carrying my stretcher onto the dock. I must have been one of the first off, because for a moment I was completely alone. I could see and hear and think.

The stretcher bearers came and put another fellow next to me. I looked at him and saw just a pile of battered clothes and a messed-up body. He was covered with blood, and the small bit of skin I could see was pale white.

It was my brother. I barely recognized him. He was unconscious, maybe close to death.

Men carried more bodies off the boat. Finally, I was lifted again and slowly walked to an ambulance. They put my brother alongside me and closed the doors.

Bill's arm had been sliced clear to the bone. That makes it sound neater than it was. To get the picture, you have to imagine something like a chain saw ripping at the skin, or maybe a lion raking its jaws across the arm. Even bandaged it was a raw pulp, his uniform a bloody rag.

We went to a nearby hospital—likely the 50th Field Hospital at Weymouth, not that far away. As I was carried out, I realized there was a crowd of people packed around the entrance.

What does a shot-up soldier look like?

Ah, I see.

Thinking back, I know that must not have been what they were thinking. It must have been something more like, *These are the brave lads who are freeing France.*

What can we do to ease their pain? To save them?

But at that moment I wasn't thinking clearly. I was confused, and hurting. I didn't think I was going to die—somehow I knew I wasn't going to die. But I didn't know much beyond that, and I was worried for my brother.

I was carried to an operating room tent. The doctors went to work. They cleaned me up pretty well and changed my bandages. The doctor or one of the nurses said something about my back being crushed.

Got that right.

They gave me more morphine, took care of my wounds. My back was immobilized; I would need further surgery in the future.

When they were done working on me, I was taken out to a ward tent. I was conscious by then; the walk seemed to take forever, my body hanging heavy against the stretcher.

They brought another man in and set him down.

It was Bill, still white as a ghost though cleaned up quite a bit. He looked more dead than alive. But he was alive, there was that. And at least so far, they hadn't amputated.

HOME FRONT

That night back in America, my wife, her family, and people all across the country tuned their radios to hear President Roosevelt give a special speech. It began simply enough, with Roosevelt's familiar voice telling them that our troops had crossed the English Channel and invaded France.

And then, Roosevelt began to pray.

"Almighty God: Our sons, pride of our Nation, this day have set upon a mighty endeavor, a struggle to preserve our Republic, our religion, and our civilization, and to set free a suffering humanity," said the president. "Lead them straight and true; give strength to their arms, stoutness to their hearts, steadfastness in their faith."

Though he didn't mention casualties or even talk about how far we'd gotten, Roosevelt didn't sugarcoat the reality. The war had a long way to go.

He asked everyone to keep praying that day, and for all the days that would follow until the war had actually been won.

The headlines in the papers the next day screamed the news:

INVASION

ALLIES LAND IN FRANCE, SMASH AHEAD; FLEET, PLANES, CHUTISTS BATTLING NAZIS

INVASION IS ON

BEACHHEADS ESTABLISHED

And so on. Pretty much every paper in the country pulled out their biggest type—they called them "wood" because they were too big to be made of the usual lead—and filled their front pages and several more with what they knew of the battles.

It wasn't much; the stories were heavily censored so that our actual positions or plans would not be given to the Nazis. There were maps, though many of them were wildly inaccurate. Even so, by piecing together the different stories over the next few days, a reader could at least get a rough idea of where we were.

Spontaneous celebrations broke out across the country. In blue-collar Washington Heights in New York City, a serviceman home on leave ran into a candy store and grabbed a small American flag. He danced outside with it, leading the neighborhood kids in an impromptu parade. On the other side of the city, in Astoria, Queens, a family began measuring the troops' progress each day by the maps that showed how close they were to Germany.

Americans did the same all across the country. For us, the war was now nearly three years old. It had been a very hard three years. Landing in northern France meant the end must be in sight.

Looking back now, most of us think of the war years as a time when everyone was working together and people were happy to make sacrifices for "our boys."

That's certainly true. But things were more complicated than that. There are parts of the picture that generally we don't see or notice. Like the strikes during the war years, which affected transport and various industries.

I don't know whether workers were justified or not in any of these cases; I haven't studied it. But I do know that a lot of us fighting on the front lines got mad when we heard about the strikes. We thought it was a hell of a thing for people to be holding out for more money when we were putting our lives on the line. The strikes may not even have been about money, but details were scant.

Rumors and emotions weren't. An artillery unit ran out of ammunition on the battlefront; rumors followed that their shortage had been caused by a strike back in the States. It wasn't true, but there were enough dark feelings around that it was easy to believe.

Why wasn't everyone back home pitching in like we were? We didn't want our parents and families and others back home to experience the horrors we were experiencing, but we did want to feel that they were doing as much as they possibly could to help us end those horrors. There were times we didn't get that feeling.

People supported the war. They loved and supported the soldiers and sailors fighting it. At the same time, they had other things to worry about, especially with their husbands or sons away.

Priorities were complicated. Not everyone wanted to fight, let alone risk death. There were manpower shortages in the army, and difficulty getting enough replacements. That could always be fixed with the draft, but the point is that looking back, a lot of times things seem simpler and maybe even better than they actually were.

I'm not complaining, but I'm a straight shooter, and that's how it was. It wasn't a nirvana of people all pulling together 100 percent of the time with never a hint of complaint. The war brought out the good in many, but it didn't make them or anyone any less human.

The soldiers knew the sacrifices that we made. We saw our friends die, sometimes next to us, sometimes in our arms.

WHAT'S MOTHER GOING TO THINK?

The next morning, Bill and I woke up around the same time. He looked over from his bed and realized I was there with him.

"What are you doing here?" he asked.

"Same thing as you."

"What's Mother going to think?" he asked.

It was a good question. She'd been notified that we'd both been wounded in Africa and Sicily. I can't imagine what she would have thought seeing the telegram about our injuries at D-Day.

His arm and his leg had been very badly damaged. A fourteen-inch slice of skin, muscle, and other tissue was missing from his leg, the bone barely intact. Shrapnel or fragments, whatever had hit him, had torn him from knee to hip. His right arm was gouged from his shoulder down to his elbow. The doctors were debating whether he would keep those limbs or not. But for now they were still there, heavily bandaged.

There's a brief account in John C. McManus's book about the 1st Division in Normandy, *The Dead and Those About to Die,* describing what had happened to Bill on the beach. As first sergeant, he'd come in with the last members of the company. He was right in the water, barely out of the landing craft. The men with him got him to shore, and there, still under fire, a medic or medics took care of him as best they could. They cleaned and wrapped his wounds, shot him with morphine. Whoever was working on him knew he was in bad shape. I doubt they knew it was my brother; I doubt they knew he was anyone but a soldier they needed to help.

At that moment, the machine guns were still firing from the

bluff. They were ripping across the nearby rocks, splintering bits of stone everywhere.

There was no cover on that splinter of beachhead. To save this man, he had to be protected.

The medic or medics working on him decided to make use of the only cover available at that moment on the beach—dead bodies.

Bill's battered comrades, likely including soldiers he had saved earlier in the war, formed a barrier around him: one last service, one last sacrifice for a comrade.

I'm not sure who worked on him. It might have been my aid man Charles Shay. He was assigned to that company. On the beach, though, at that moment, it could have been any of a half dozen guys, even someone from the 29th Division or a later wave.

Whoever it was, they took great care of him. But then, they did that for everyone.

Bill and I talked a little bit that morning until the hospital staff came and gave us some food. Later, they sent us north by train to Cheltenham and the 110th General Hospital, a large facility set up around an older English hospital and manned by the American army.

I remember that train ride. It jerked and bumped and rocked and shuddered every few seconds, the tracks barely in line. The springs on the cars did more to amplify the shocks than absorb them.

Every time we hit the slightest bump, Bill would cry out in pain.

Our wounds were different and so were our prognoses. We were separated, with Bill put into a ward where he could get more intensive care. It took a few days, but when I was feeling more myself, I asked the staff if I could see him. They pushed me down in a wheelchair to his bed.

"Sending me home," he told me.

"Good."

My true thoughts were more complicated than that. Sending him home meant they were sending him somewhere to get more care. It was good that he was stable enough to move; bad that he needed that level of care.

The doctors were still debating about his arm and leg. But for now they were there, tucked tight in bandages.

"We'll be okay," he said.

"I know we will."

BATTLING ON

Young people sometimes think that the war ended when the sun set on D-Day, or that things were all easy from there.

Not true.

The first-day objectives—in other words, how far the Allied units were supposed to go—were entirely too optimistic; very few units got even close. The Germans recovered from their initial confusion and began organizing counterattacks and better defenses. Rommel arrived on the scene. Though they were outnumbered, the Germans were still extremely able fighters. They were able to prevent the British from taking Caen, and continued to do so for weeks.

The two American sectors inland from Omaha and Utah (the technical term is "lodgment") remained just that—two separate sectors—for several days. They weren't joined solidly together for another week, until the 175th Regiment from the 29th Division on one side and the 101st Airborne on the other repelled German counterattacks around Carentan. In the meantime, the bulk of the units in each section advanced against different targets.

Following their landings, the troops surging from Utah started for Cherbourg. They encountered heavy resistance. Flooded fields, narrow, unimproved roads, and fierce German fighting slowed the advance to a crawl.

The generals changed our strategy. The troops would still take the port, but first we'd cut off the Cotentin Peninsula, moving across to the west coast of France. This, too, proved difficult, but by June 17 a firm line had been drawn and Cherbourg was cut off from further help.

The German defenders there held out until June 26. When American troops finally reached the docks, they found them destroyed. It would take over a month before they could be used.

At Omaha, the 16th Infantry's 3rd Battalion joined with a battalion from the 26th Regiment to strike southeastward and take the high ground west and southwest of Port-en-Bessin on D-Day+1, while the 1st and 2d Battalions of the 16th cleaned up the resistance inside the captured area. We took a stab at trapping Germans between us and the British, but the Germans were too strong and our units were not yet organized and rested enough to pull it off.

Regrouping, the 18th and 26th were part of an attack intended to reach Saint-Lô, a crossroads city ten miles from the shore. The offensive reached Couvains, about a ten-minute motorcycle ride from Saint-Lô, but went no farther.

The Battle for Normandy had reached a new phase. The question was no longer whether we could get into France, or even off the beaches. The question was, how would we get enough momentum to break the German army down and liberate the rest of the country?

As my evacuation demonstrated, the Medical Department followed the plan that had been set up in the months before. By D-Day+3, the military term for the third day after the landings, planes took over most of the transport of the wounded back to England, greatly shortening the amount of time it took. The next day, the first evacuation hospital was set up in France; patients were being operated on there within two hours. Not only were doctors now taking care of wounded GIs close to the front, but army nurses were there to help as well.

It was later estimated that the medical department handled 76,000 cases in two months during the battle for Normandy. About 3 percent died, a large improvement over World War I.

One last grim statistic: it's said that in the first hour on the beach, 30 percent of the men in the Big Red One were either killed or wounded.

As the American army built up in France, it found itself bottled into a small area of Normandy. As the days passed, many of the war planners worried that the battle would become static, horribly similar to the trench warfare of World War I, a back-and-forth slaughterfest.

Our guys tried to break through the German lines several times.

We had a lot of things going for us. Air superiority was one. Our planes were often overhead, dropping bombs or using their machine guns on German targets. Nazi planes that tried to attack ground units usually were intercepted and often shot down. We had a top-secret intelligence system that broke the German encryptions; this allowed the commanders to know many of the German movements. And while we didn't have nearly the amount of supplies and equipment we wanted, more was arriving each day.

But there were advantages to being on the defensive, and the Germans played them well. The biggest was the geography. Called the *bocage,* the area was broken into many small squares of fields separated from each other by hedgerows.

An American hearing the word "hedge" may think of a neat row of shrubs about knee high that need clipping every month or so to make the front lawn look pretty. The hedgerows in Normandy were masses of dirt topped by thick brush and sometimes trees. They provided perfect cover. A handful of German soldiers could set up a pair of machine guns in one and hold off a platoon for hours before quietly slipping away.

Once we took one field, we'd have to fight our way through the next. And then the next. And the next. Normandy wasn't a checkerboard of hedgerows; it was a universe of them.

It took a while to figure out the solution: rather than coming in the front door, bulldoze your way through the side.

With a tank.

Some enterprising GI—there are different contenders for the honor of having come up with the idea—welded a pair of metal prongs to the front of an M4, making the tank look like a rhinoceros. The crew then went for a spin around the nearest Nazi-held field.

Not so much around as through. The rhino charged bulldozer-like into the side of a hedgerow, cutting an entrance that the troops supporting the tank could rush through. Once inside the field, the tank used big guns and heavy machine guns against the German machine guns. It was an unfair fight, something most soldiers like.

The idea quickly caught on, and rhinos became the rage throughout Normandy. Unit after unit improvised with what they had, adapting their tactics to suit the situation.

It was still slow work. Saint-Lô, which had been only a few miles from our front line for a month, didn't fall until July 19. Even with the city taken, our troops were still bottled up.

The breakthrough finally came on July 26, when after an immense, tactical attack by heavy bombers and a massive push by the armies in the area, the 1st Division threw itself against the German defenders north of Le Mesnil-Hermanses. The Germans resisted fiercely, but in the morning, they were gone, mostly dead; our guys cleared the mines they'd laid and cut a strategic hole in the defenses.

Other units did the same throughout the area. Suddenly northern France was wide open. Their line punctured, the Germans found themselves so vulnerable that they had to retreat, and had to move so quickly that they couldn't regroup.

Troops from what was now the First Army raced after them. Others poured in from England and the U.S. Among these newcomers was the Third Army under George Patton. The general was back in Eisenhower's good graces, though he and his army were now under Bradley as part of the 12th Army Group, rather than vice versa.

Over the next several weeks, Patton would lead his army in a superb drive to Germany and the Rhine. But unit pride forces me to say that the Big Red One got there first.

ROME, RUSSIA, AND JAPAN

France was not the only place we were fighting.

Three days before we landed at Normandy, American troops entered Rome.

The Allies had been working their way up the "boot" of Italy since landing there in September 1943, shortly after Sicily was occupied. The Italians had tried arresting Mussolini and surrendering to us; Hitler squashed that by sending special operations troops to rescue Mussolini from prison. He installed Il Duce as his puppet while German troops took over defense of the country.

The rugged hills and mountains made it extremely hard for the Americans trying to advance there. Big advances were few. The slow progress may have influenced public opinion, which never seemed to give the troops in Italy as much attention as we got. After the war, and even during it, a lot of fellows complained that they were overlooked.

There were over a half million troops in Italy at the beginning of June 1944. Driving out the Germans would eventually cost 40,455 lives, including those killed on Sicily. Those men weren't a sideshow, and they certainly deserve our respect and thanks today.

Out in the Pacific theater, marines and army soldiers landed on Saipan the week after we hit the beaches. Located north of Guam in the Philippine Sea, the island's location was strategic not only for the navy, but for the army air corps, which was getting ready to introduce a new weapon into the fight against Japan—the B-29.

The four-engine, high-altitude bomber would use the island as a base for launching long-range attacks on Japan some fifteen hundred miles away. Those attacks would eventually include fire-bombing Tokyo and the two atomic weapons that ended the war over a year later.

On June 22, more than two million Soviet Union soldiers launched a massive attack against the German troops still occupying their country. The location of the attack, in the center of the German line, took the German forces by surprise; they were expecting an offensive farther north.

By August, the Russians would be deep inside of Poland. The Axis was running out of time.

HOSTAGES OF WAR

Reducing war to victories and defeats, talking about armies moving back and forth—at times we might think this is all happening on an open plain, a vast empty space like a chessboard.

But most of these battles took place where people lived. Some had fled before the armies came, often escaping with nothing more than a few clothes, cast out to wander as strangers in a country that was no longer their own.

Those were the lucky ones.

Others stayed, hiding in basements during the fighting, coming out only when necessary to find water or food. What followed when the guns stopped shooting could be worse.

In France, families who survived the German invasion in 1940 and lived through the Occupation were always one phone call away from being arrested. A jealous neighbor could easily make up a pretext for the Gestapo to investigate. Even the slightest inquiry could lead to torture or a death camp.

Betrayal wasn't a necessary prelude to death. In many of the towns and villages throughout France, the occupiers designated prominent citizens as, in effect, hostages. If there was trouble, reprisals would be taken; these people would be killed.

The more popular—the kinder, the more valuable, the more important—a person was, the better a hostage he or she made.

Random monuments to these people dot the Normandy countryside. Buildings and squares remember their sacrifices simply and eloquently. A rebuilt church has the names of soldiers on one side of a monument, murdered citizens' names on another. An obelisk at a farm crossroads declares, *He died for his duty.*

As the Allies broke into France, much of what was in our way was turned to rubble. We broke the German defense at Cobra by obliterating not only the defenders and their weapons, but the buildings and everything else. It was an unfortunate necessity.

That is the worst of the many terrible parts of war. Things that we do not want to do become the things we must do to survive.

And then there are the things some do that can never be excused.

I had been in recuperation for only a short time when the so-called Vengeance weapons began falling in southern England. These were the V-1 and V-2 rockets. Not only did they inflict serious damage and kill many civilians, but they reminded people of the terrible air raids earlier in the war, the Blitz.

I was on a rare visit to London when air raid sirens sounded. We descended into the city's subway system, the Tube. The descent seemed endless, especially as I was still having trouble walking.

As the crowd moved steadily downward, I wondered if we were ever going to reach the bottom. The people around me didn't panic; there was no pushing or shoving, just steady walking. A few detoured around me, but mostly we moved as one.

We stayed belowground for what seemed like hours, though it was probably not much more than twenty minutes. Occasionally someone might make a joke, but for the most part it was silent, and grim.

If there was any damage from the raid, I didn't see it. What I noticed was a grim determination mixed with resignation. *How long must we go through this?*

And maybe sidelong glances, wondering why soldiers weren't doing more to end it.

SHIPPED OUT

They flew Bill home after only a few days at the hospital. He was still in serious condition and would undergo a number of operations, including plastic and reconstructive surgery, but by then I gathered that the doctors thought they could save his leg and arm.

Nothing I said could assure him, though. And I understood. Having been a medic, he must have known that keeping his limbs and even his life might not be possible. Assuming he made it—not an easy assumption—he faced many operations.

One thing definitely helped him: his wife happened to be a nurse at the hospital in Massachusetts where they were sending him.

That turned out to be the best therapy possible. She made sure he got great care. You can't do better than having your wife as your patient advocate, especially when she's a nurse on staff.

My time at the 110th General Hospital was just beginning. The wounds in my leg and arm and the other little nicks and cuts were washed and sewn up. The doctors fused my backbone in the area of the fourth and fifth vertebrae and did some related work there.

It was the state of the art for dealing with broken backs at the time, and the prognosis was decent. Fusing my back like that meant I wouldn't be able to bend over quite as far as I had before—I couldn't tie my shoelaces—but it was far better than not being able to walk at all, or, worse, having been paralyzed and unable to feel anything in my extremities.

My leg would turn out to be the bigger complication. The outer skin and layers healed, but the wound below those levels got infected. Dealing with the infection would take several months.

Long term, the fusion caused arthritis, which I feel today, especially on certain days. But otherwise, the lasting effects of my injuries are lines and marks on my skin. Knife wounds heal better than fragment wounds; they're straight and clean compared to jagged and usually dirty. Because of that, the bayonet's mark is now a very faint scar; the one at my elbow is more prominent. They are my mementos of the war, reminders that will never leave me.

I'd lost most of my other mementos, including a very sweet German pearl-handled pistol I'd picked up from a dead German officer. I'd made the mistake of putting them and a lot of my clothes with our Jeep, thinking this way I would have my things very soon after we landed.

That would have worked had the ship with our Jeep not been sunk. Some other items that had been stored in a barracks bag—a large duffle bag typically shipped well after an invasion—showed up a very long time later. Among them was the bayonet that had sliced my hand.

I did have two things besides my scars that would remind me always of the war—the pipe the German observation pilot had given me as he died, and the photo of the man who tried to kill me.

Both had been in my medical bag, which was still with me when I passed out on the beach.

As I got stronger, I was moved to a rehabilitation ward. Soon they organized the ward into a company and put me in charge, making me the *acting* first sergeant of the 110th Rehabilitation Company. We were a group of men recovering from various injuries that left us, in layman's terms, walking wounded, but too weak to pull regular duty.

I tried to lift spirits when I could. A lot of the guys were pretty shot up; just paying attention to them seemed to help, but I know it wasn't a miracle cure.

I said I was "acting" first sergeant. To make the position permanent, I would have needed to extend my service and reenlist. I thought about it—there'd be a raise and a promotion.

A recent change had made it possible for non-doctors to hold officer rank in the medical department—did I want to apply?

I should apply, said some of the officers, who even went to the trouble of writing recommendations, hoping to persuade me to use them.

My leg healed much more slowly than either I or the doctors wanted because of the infection. My back was even more complicated. It hurt when I walked; it hurt when I rested. It hurt all the time. I refused to use a cane, though. It was going to hurt one way or the other, so why make things worse with a cane? I thought that if I kept going, slowly building back my strength, eventually I'd get better.

Some days, "eventually" seemed far away.

In early July, I received a letter from Major Tegtmeyer, who was head of our regiment's Medical Detachment. Tegtmeyer told me the unit had been awarded the Presidential Unit Citation—a high honor—for our work on the beach. He added a personal, handwritten note at the bottom: "You are also entitled to the Bronze Star."

Not too long after that, an officer pinned a Silver Star on my uniform.

Ordinarily, you get a citation or other paperwork that formally records the reason for the medal. In this case, I didn't get any paperwork; without the commendation, it's not really official. When I felt better, I wrote to our division to try and track it down.

Five months later, a four-page handwritten note from the major arrived.

"You have a Silver Star coming," he assured me, apparently referring to the paperwork. I had been nominated for it by Colonel Herbert Hicks, our 2nd Battalion commander. Hicks had been on the beach with us, and had been awarded the DSC for bravery there. "I'll take care of it," said Major Tegtmeyer.

I don't know what happened after that, but the paperwork I was supposed to get never arrived. Decades later, when I finally thought of looking for it, no record could be found—nor could they find the details of the one I received in Sicily. Of course, by that time, Hicks and Tegtmeyer had passed on. My enlistment record lists only the first Silver Star I earned in Africa.

So the official commendations and their whereabouts are a mystery, and will always be. One more thing I have in common with many of the men who were on Omaha Beach that day.

If the clerk who should have forwarded me a copy of the paperwork for 1st Division got distracted right after Tegtmeyer wrote, it's understandable. On December 16, two days after the major wrote that letter, the Germans launched a massive offensive in the Ardennes. We remember it as the Battle of the Bulge.

The Big Red One had been worn down after taking Aachen on the border in Germany and then slugging its way through the Hürtgen Forest to the Ruhr River. Exhausted from months of combat, the division was cycled off the line for rest and regrouping on December 7—only to be called back barely a week later as the Germans advanced.

The surprise attack caught the U.S. off guard. At first the Germans made impressive gains in what had been a quiet, weakly defended sector. But resistance quickly stiffened, and as reinforcements poured in, the Germans found themselves in danger of being trapped. They gave up ground slowly but surely. Our guys beat them back to the Rhine, then crossed at Remagen, beating Patton again and setting the stage for the endgame in Germany.

The Regiment Medical Detachment had held my position open since I got hurt, but by then it was obvious to me that I wouldn't be able to join them, or any combat unit, in the near future. I still couldn't walk very well, and I'd lost a lot of weight and strength.

That was my reality. And if I wasn't going back to my men at the Big Red One, I wasn't staying on here, either. The offers of promotion and the recommendations were all very flattering, but I wanted to go home.

I hadn't seen my wife in three years. I had never met my son.

I'd done a lot in the war. It was time to let others take over.

"DADDY"

I boarded the *Queen Mary* in January, around the time when the 1st Division was rolling the Germans back to close the Bulge.

Before the war, the *Queen Mary* had been one of the most famous luxury cruise ships in the world. There were two swimming pools and a grand ballroom. Gentlemen in first class couldn't enter the dining room unless they were wearing their evening clothes, which weren't pajamas but bow ties and jackets.

It wasn't the sort of ship a GI from Alabama would ordinarily sail on, that's for sure.

The liner had been converted into a troopship after the war started, and sailed everywhere from Sydney, Australia, to Bombay, India, before settling in as an American-run transport going back and forth between the U.S. and England. Painted gray, she remained one of the fastest liners on the ocean.

The drab paint scheme earned her a new nickname: the Gray Ghost. Inside you could see traces of her former beauty, though these were often hidden behind something more utilitarian. There was a beautiful set of stairs leading down to the first-class dining room, now a mess hall for the enlisted men. Bunks now filled the cabins—every cabin, not just the third class. Hammocks were hung in the theater. Any area that could be used for passengers was taken with some sort of sleeping gear.

Though I didn't work in the ship's well-equipped hospital this voyage, I was given a cabin to share near it. That spared me having to fight my way below for meals and whatnot; we were fed there, in relative luxury.

I'm not sure what I did to earn the privilege. Possibly they thought I could take care of the other fellows if their wounds or ailments gave them trouble. I didn't ask; I'd long ago realized you don't question certain things in the army.

After we docked in New York, I was taken to a hospital in Manhattan for evaluation. It was a standard stop; doctors would check us over and then send us to another facility to continue rehabbing.

I was hurting, but could get around on my own—slowly but surely. Soon after arriving, I went down to the lobby to use the telephone. So did half the hospital, or at least it seemed that way. I must have waited on line several hours before it was my turn.

I called my parents. After a very brief conversation, telling them that I was okay and was in the States, I asked where Estelle was.

"Her folks."

"Can you give me the number?" I asked.

I dialed pretty quickly, nervous all of a sudden; she knew I was on my way, but it had been so long since we talked. And no matter how sure you are of someone and their love for you, there's always some little piece of nagging doubt.

Does she really want me back?

Estelle's sister Becky, our old chaperone, answered the phone.

"Is Estelle there?" I asked.

"Who is this?"

"It's Ray."

"Ray's home!" Becky shouted. "Ray's home!"

Estelle was overjoyed. We talked for quite a while. I told her about the injuries, how badly I wanted to see her, how I was going to get better in no time.

"Can I come see you?" she asked.

"No," I said. "They're going to move us in a day or so."

I told her I would call her as soon as I was moved.

The next day, an aide brought a woman and a young boy to see me.

It wasn't my wife. It was the wife of a friend. I recognized her instantly, though I hadn't seen her in over three years.

"Why did you let him die?" she said before I could open my mouth. "Why did you let him die?"

"Who?" I asked.

"Puckett," she answered, naming a mutual friend who was part of our medical section.

I hadn't even known he was dead.

She began to cry. I hadn't even known the man was dead—I found out much later that he had been killed long after Normandy, while I was recuperating. But that was how close we all were before the war began—families.

She began pounding my chest. I said something useless to get through the moment. Slowly, more words came. I told her I didn't know that he'd died, and that everyone on the team would have done everything they could.

She collapsed, crying.

"I know. I know," she said between sobs. "I'm sorry."

I told her I was sorry, too.

I was assigned to Finney General Hospital in Thomasville, Georgia. I had one mission—get better. My therapy largely consisted of rest.

A few days later, Estelle and my son drove from North Carolina to see me.

"Daddy!" he said, as soon as he saw me. "Daddy."

I bent to him. "How?"

Estelle explained that they had hung my picture over his crib and then his bed from the time he was born.

"Daddy."

The word melted me. I was home, and now finally the war would be over for me.

TWELVE

The Rewards of Peace

GETTING BETTER

He's there, in front of me.

He has a bayonet!

Explosions and thunder and a roar so loud my skull vibrates.

That man is drowning!

I dodge into darkness, but someone follows. I'm in Sicily, then Africa, on the hill with the German, the bayonet slicing my hand. I plunge into a different landscape entirely—Battery Park, New York City, before the war, a fistfight in the men's room of a bar. I find myself not in the bar but beside a wrecked airplane as a cadaver hands me something, a pipe . . .

But I'm underwater, held there, trapped, about to die.

Breathe! . . .

Finally, I wake up. I'm in Florida, still in the army, convalescing. I am in absolutely no danger, but every part of my body trembles with fear.

I'd thought the war was over for me. I was only partly right.

No one was shooting at me anymore. No bombs were exploding. But the war stayed with me, and stays with me still. The nightmares I had as I recovered came less and less often as the years passed, but even now I still occasionally have one, or a piece of one.

I try not to go to sleep with bad thoughts on my mind; it helps.

I have more coherent memories and thoughts as well. Not just of the man who tried to kill me, or the medics I worked with who died in the war, or even of the thousands of soldiers my guys and I worked on, the men we saved. I think sometimes of the lives that came after, of the children and the children's children of the men we saved, people who would not have been here but for some mixture of bad and good luck that made my presence not only necessary but provident.

I have been blessed.

I don't know if the disjointed dreams and nightmares I had in the hospital were symptoms of post-traumatic stress. We didn't have that diagnosis in those days; we knew the war had a psychological effect on people, but we thought of it differently than we do today. Even the doctors had a different attitude than is common these days.

I do know a lot of guys had trouble adjusting after the war. And I know that I changed, physically and mentally, because of the war.

I'd lost a lot of weight; I was down to 136 pounds from over 170.

And I was more nervous, more likely to worry than to take life as it came.

Some of that anxiety was justified as my time in the army came to a close. I didn't know how I was going to support my family once I left. I was still recovering from my wounds. I had all that on me, no job, and no place where we could live.

But I did know I had to just go ahead and get on with my life. I imagine it was like the old days when the pioneers set out with their covered wagons. They had no real knowledge about where they were going, or a deep understanding of the trials and trouble they faced.

Most thought it would be easy: just drive out west a few weeks, find a piece of land, live happily ever after.

Boy, were they wrong.

I wasn't that naïve. But I was in a kind of similar situation. I had to just grit my teeth and go on.

Which is what I did.

VE DAY

After Georgia, I was sent to another hospital in Florida. We found a cheap and tiny place to rent in Daytona Beach. Not only was it tiny; it had been a chicken coop before we moved in. But some of the guys from the post went in and cleaned that all up, even whitewashed the walls. There was no electricity or running water; to use the bathroom we had to go into the landlady's house, which fortunately was right next door. There was no door to close when we moved in; I had to buy one on my own. Eventually we ran an extension cord over for light.

And that's how we lived for twenty dollars a month until my discharge came through in June 1945.

I'll say this for my landlady—she was a nice woman, and didn't mind that we had a child, unlike many others. The town was full of signs saying soldiers with kids need not apply. People were happy to jack up the rent and take our money; in many cases that was the limit of their generosity.

Germany had surrendered by then, formally exiting the war on May 8. Hitler committed suicide in his bunker on April 30; a few days earlier, Mussolini had been killed and hung up on a meat hook in a public square by Italian partisans.

There were wild celebrations across America. I don't remember joining any, though obviously I would have been pleased and thankful that the war was over. I had other things on my mind.

My brother had been discharged as well. He'd kept his leg and arm. I knew he would. He had that kind of will.

He'd found work with an electrical company in Massachusetts and suggested I could, too. That sounded like a good idea. Estelle and I took our son and a newly acquired dog, bought a cheap car and a trailer, and set out for the North. We had a $275 stake—all the money left from my discharge and savings.

I didn't know my wife, and she didn't know me. It wasn't just that we hadn't seen each other in three years, or that we'd gotten married very young.

We'd both changed. Me, because of the war. And her because of the war as well.

Less directly, of course. But the war had kept her husband away, and left her to raise an infant without him. The fact that this happened to many other women across the country didn't make it any easier, just as the fact that there were a million other guys in the army didn't really help me either.

She was patient as well as loving. Those two things don't always go together, but in Estelle they did. And that made our marriage and life together thrive.

She would never, ever argue with me; she never yelled at me that I can remember. She never said a bad word to me, called me any kind of name, or told me she was unhappy about things. If she'd been someone else, I might not have made it.

We settled in Waltham, Massachusetts. I got a job as an electrician's helper; at night I'd go to school to get certifications that would allow me to do bigger jobs and earn more money. There was a drugstore across from where we had a little apartment; I took a part-time job cleaning it, washing and wiping the floors and counters and everything for a dollar. On Saturdays I worked for a guy who had some odd jobs and needed wood to heat his house, which in a funny way brought me full circle back to the wood business. I'd spend a few hours cutting down trees and splitting them into firewood.

I took any work I could find. If I wasn't doing anything, fifty cents was more than I had in my pocket.

It took a little while for me to get the knowledge and experience I needed, but by the early 1950s I was ready to go out on my own in the electrical field. I moved out to Westwood, Massachusetts, hired a couple of guys, bought a bucket truck or two, and went to work.

My early customers included a large dairy; I was back on the farm, in a way. The business kept expanding. Besides servicing factories and warehouses for companies like Coca-Cola, we hired out to Boston Edison to help get electricity back on following the storms. Nothing like being in a war, but pretty demanding and potentially dangerous.

The business grew very successful; so good, in fact, that in the 1960s I bought out two competitors and eventually found myself running a good-sized company. There was time for fun, too—snowmobiling in the winter, for example. Estelle and I and Arthur and Linda had a blast.

Linda—my daughter. She was born in 1951 and, as the saying goes, was the apple of her father's eye. I was able to spend time with the kids as they grew up, and found myself volunteering in the Scouts with my son and helping establish the first girls' softball league in our local area.

There's a bit of me that's an inventor and tinkerer, I guess, and as I was working in the field, I realized that the heat generated by large electrical motors could be put to use. The idea grew and eventually I refined it, hired someone to help me, and we came up with a device that reclaimed heat and used it for different applications; you could heat the factory or offices with energy that otherwise would be wasted.

For complicated reasons, we weren't able to patent the device, but we were quite successful with it. We shipped units all across the country, and even the world.

My brother, in the meantime, decided he couldn't stand the cold—not that I blame him. He moved out to California, started his own business there, and was quite successful. And warm.

LOVE, SORROW, AND FAME OF A SORT

In the winter of 1981, Estelle had not been feeling well and eventually made an appointment to see a doctor. I waited with my daughter while Estelle went into the examining room. It took an awfully long time—you know how doctors can be.

Finally the door opened and the doctor came out.

"Your wife has lung cancer," he told us. "We're going to try to operate."

Estelle was her usual cheerful self when she came out a short time later. She went to the hospital and they removed a lung. The next day the doctor told me that the cancer had spread.

They tried chemo. She lost her hair and suffered greatly. She would stay in the hospital overnight, coming home the next day drained.

Treatments in a Boston hospital didn't help; it seemed as if they did more damage than the cancer.

Almost a year to the day after she was diagnosed, Estelle fell into a coma in the hospital where she was being treated. It was near our fortieth wedding anniversary.

On the day of our anniversary, my son and daughter came with me and a pair of friends to the hospital. We had a small bottle of wine, a cake, and roses.

Estelle was unconscious the whole time, and died two days later.

She was a beautiful, loving woman. Her only flaw: she smoked cigarettes, and it killed her.

I filled my time with work, the Lions Club, a few other business and community things. But there is a hole in a man's soul if not his heart when he doesn't have a true companion.

It was my daughter Linda who knew that I was drifting.

"Mom would not want you to be alone," she said. "She'd want you to be happy. Start dating."

She played matchmaker, recommending I ask out a lovely lady who worked in the office of one of my companies.

I couldn't do that, I told her. She worked for me.

Eventually, my daughter persuaded me.

Looking back now, I suppose I was very awkward about it, or careful, or both.

Being with her, though, was very easy. And in May 1983, I married Barbara Mahan.

I'm not exaggerating when I say her love has kept me alive.

Time passes without us having much say in it, doesn't it? And with that come passages.

My brother Bill passed away in 2010. We spoke for a long time the day before he died. He'd lived a good life, he told me; he was ready to move on to eternity. Thinking back on it, it was a sad conversation, and yet what he said was so characteristic of him that I can only admire his presence of mind at the time. He was both philosophical and practical throughout his life; he was clear-eyed to the end.

I'd retired by then, having sold my businesses when I turned sixty-five. A few years later, Barbara and I moved to North Carolina to enjoy our retirement.

People think that life slows down once you retire; you can't prove it by me. I seem to have more commitments these days than I did running two firms.

Not that I'm complaining.

I've had a few hobbies over the years. Restoring old Mercedes—I loved the 280 SLs especially—was a lot of fun. One of the most rewarding things I've done in my golden years was connecting with veterans of the 16th Regiment and the 1st Division. I have been honored as a distinguished member of the regiment's association, as well as a member of the Society of the First Infantry Division. I've even been made an honorary tank commander—quite a privilege for a medic.

The reunions and staff rides across Europe are a lot of fun. I've had the privilege of visiting the Normandy battlefields several times. In 2018, the village of Colleville-sur-Mer, where I landed on D-Day, dedicated a monument to the medics and myself at what is now known as "Ray's Rock." It was an honor and a privilege to be at that ceremony. I was very fortunate that a number of my friends and family could attend as well. The French have always been tremendously gracious to me and other veterans, always expressing their gratitude for our role in liberating their country.

Having survived the war and lived so long, I've become a point of contact with the past. Generals and others have sought me out, not just to thank me for my service, but through me to thank all who served and sacrificed in that war. I think I've met every 1st Division commander since 1940 and gotten to know most of them well. Same for the Army chief of staff and any number of other high-ranking officers. I've even met the queen of England, a rare treat for a kid from Alabama.

While there have been a number of news stories and such about me or quoting me on D-Day and the war, I'm always a little surprised that people know my name or recognize my face. Odd things happen—I turn up on the internet; a hotel where I'm staying finds out and puts my name on its sign to welcome me.

It's always a little touching, not to mention a surprise. I try not to let it go to my head, knowing it's really my companions and generation that are being recognized.

Unfortunately, there are many fewer of us these days. I know only a handful of other men still alive who were on the beaches on June 6. Soon there won't be any.

I worry that the connection to the past, to the values that put us on that beach and saw us through that terrible day—values that took us from Africa to Italy to France and beyond—will weaken and die when we are gone.

But then I remember that they are enduring values that we didn't invent. They have withstood a revolution, a civil war, untold conflict. The human vessels they are contained in are imperfect; we struggle to improve ourselves and our relationships to others and the world. But the values themselves will endure, and our connection to them will as well, as long as we remember and honor the sacrifices others made in their name.

THIRTEEN

Legacy

I was born in a place of rich soil and richer possibility, at a time far different from the present, yet recognizable by the threads that have led us here. I was endowed with great wealth, but not money. It was a simpler world in the ways that are often spoken of, but I and my generation faced the same complex truths and harsh realities mankind has wrestled with since Adam and Eve. As many before and since, I discovered that evil can be an overwhelming force in the world, that fear and courage can be close companions, and that wanting to do the right thing is no guarantee that you will succeed.

I also discovered that loss can be overcome, that love can survive separation, and that saving a man's life is its own reward.

I learned what it feels like to kill. I learned what it's like to see the last breath of a close friend.

Some things I saw and learned cannot be fully described.

What words can explain the way a bullet spins through a man's head, or the rattle of a leg slapped free of its body? But there are many other things I saw that all of us, being human, can understand and feel—the grin on your child's face Christmas morning, the smooth curve of your spouse's body as she bends to set the table for dinner.

Every man's life takes its own course, tracing a path like a solitary raindrop, sliding down the pavement. But that path can be seen as part of a much larger whole, a river perhaps, running to the sea. And so my story, with all its specific twists and turns, its tragedies and happiness, is not just my story, or even the tale of my generation; it is the story of America and the world. The late 1930s and early 1940s saw mankind struggle to answer a very basic question: Should good or evil prevail? But day to day, that larger question was far in the background, pushed away by more immediate ones, like, *Will I live today?*

There are not many of us left who walked the paths I walked. Last year at the 1st Division reunion, there were only two of us who had gone in on D-Day. I was the only man there who'd made all three landings—North Africa, Sicily, and Normandy.

I'm telling my story in these pages not for my glory or even that of my fellow soldiers, but for the future, for my grandchildren's grandchildren, so they may understand that the struggles they face are not insurmountable. They will live in a far different time than I, but we will have much in common. They, too, will be tested and tempted; I pray that my story will make some small contribution to the ultimate triumph I know they will achieve.

Sometimes when I'm asked about the awards and such, I honestly don't remember the details. Often I'll leave one of the medals out.

I don't mean that they're not important, or that I don't appreciate receiving them. It's that what I had to do at the time was help or save someone, and I was focused on that, not getting a medal or an award, or even a thank-you. I had to do my job. Even more, I had to help that guy in front of me.

We dodged bullets and shells all the time. On the beach in Normandy, in the mountains of Sicily, every time we were under fire—every man was a hero. We didn't stop to calculate: *If I do X, then I will receive Y.*

Run into this minefield, receive a Silver Star.

Pull this soldier out of the surf, get a Distinguished Service Cross. Medal.

War isn't like that, and we weren't like that. Plenty of times I did things just as brave or foolhardy as the things I was honored for without getting a medal. Plenty of guys were far braver than me, and all they got was an honorable discharge.

The medals are important. They look nice in the display in the corner of the room. But the real place honor lives is in the hearts of other men and their families, the guys you saved, the families and generations that exist because of that one act of courage—or risk, or foolishness, or even temporary insanity. No general pins anything on your chest close in value to the legacy of the man you helped, his sons and daughters, his grandchildren.

Cherish the honors and awards, but realize the true measure of a man lies somewhere else.

Collaborator's Note:

Bringing Ray's Story to the Page

I'd been working with Ray on this book for several months, talking on the phone, following him to Europe, and interviewing him at his home, when late one afternoon we took a break to take some photographs of memorabilia. As we were setting up, I happened across an official-looking memorandum typewritten on a plain piece of paper. Not sure what it was, I was about to set it aside when a few random words caught my eye.

It was the citation for his first (as far as I can tell) Silver Star, awarded in Africa for rescuing men being overrun by the Germans.

Remarkable in itself, surely, but to me even more remarkable was the fact that in all the time we'd been working on the book, Ray hadn't mentioned this incident.

"Oh, yeah," he said, or words to that effect. "The Germans advanced and it was a little tricky to get the guys out. I had to go back two or three times . . ."

Why hadn't he mentioned it before, I asked.

"It wasn't anything special."

To say that the heroes of World War II are humble has become both a meme and a cliché over the years, but it's also true in many cases, and certainly this one. While Ray has been an active member of veterans' groups associated with the 16th Infantry and the 1st Division for many years, it wasn't until the last few that he decided to record his story for his family; our book was even more recent.

This was despite the fact that literally thousands of people have urged him to write a book over the years. He didn't because he didn't think his story was worth telling. The only thing that persuaded him was the fact that, at ninety-eight years old, he is one of the last survivors not only of D-Day but of America's World War II European campaign. While there are plenty of books on World War II, making a connection with history often requires personal contact, and Ray rightfully feared that as veterans of the war passed away, so too would the memory of their struggle.

Human memory is, after all, a very fragile thing. And unfortunately as direct memory of a thing is lost, too often the lessons that it taught are lost as well. In the case of World War II, without doubt the most costly, ferocious, and tragic worldwide conflict humans have ever known, that loss could easily prove catastrophic. Both Ray and I fervently hope that adding his story to the collective memory will help ensure that never happens.

Memory being fragile, both of us worried that his memory of events might not be accurate. It was my job to both prompt and provide context to his specific memories, and also to fact-check to the extent possible. He did his part as well, referring when he could to documents, photos, and notes that he'd accumulated over the years. During some of our formal interview sessions, Ray sat with a history of the 16th Regiment in his lap, occasionally dipping in to help jump-start his reminiscences, or occasionally place them into a specific time and place.

I must say, though, that his memory of the war was extremely sharp, and I'm not qualifying that by mentioning his age. Naturally, every participant has a different perspective; I've seen cases where a half dozen after-action reports on the same battle contained accounts so diverse that one had to wonder if all the men were in the same battle—and this was from soldiers barely ten yards from each other for its duration. So if there is a caveat to this book, it is that it is Ray's personal perspective. As helpful as the sources listed in the appendix were, these are Ray's memories, and his alone.

Humility has been an important hallmark of Ray's life and personality, as it was for many soldiers of that generation. It was therefore easy to accept and even to explain. A more difficult quandary was this:

How did a young man who'd never finished high school, who had been raised in small, rural communities, and who had never held a very important let alone well-paying job, become in a few short months not only a hero, but a bona fide leader in the U.S. Army?

It may well be that becoming a hero was simply natural and innate, a matter of first being raised to help and care for others, and then being in the right place at the right time—if a minefield can ever be described as the right place at any time.

I think there is a great deal of truth when Ray says, as he often does, *I did what I thought I should.* He followed his impulses and didn't stop to question them. Most times, though not always, he wasn't scared. To hear him tell it, he wasn't anything at all. He acted, rather than thought or felt.

So maybe being a hero was "just" a natural part of him and the way he was raised.

But the leadership part—surely that is something that isn't innate. There are programs and books and countless websites devoted to training one's self to be a leader. Ray, and I would guess the vast majority of people who found themselves thrust into leadership positions during World War II, never took any of those courses. But the army recognized his abilities and his skill at sharing those abilities with others. He treated the men he worked alongside of with deep respect, getting to know them, assessing what they could do, and never asking more from them than they were capable, or than he would himself do.

Where precisely those qualities came from remains, to me at least, a great mystery. I suspect that they were always there, nurtured by his experiences growing up, and able to fully blossom once he was part of the army. Hardship and sacrifice were not obstacles, but rather the very things that made those qualities grow.

If that was true for Ray, then by extension it must have been true for his entire generation of men and women. There are many reasons for America's prosperity in the postwar years; the leadership abilities and the caring for others that people nurtured through the hard times of the Depression and the world war are surely the least valued and understood. But they deserve credit.

This is not to simply restate the cliché about the "Greatest Generation," a name that papers over not only many failings but also a large part of the struggle and the triumphs. Nor is it to look at the past with rose-colored glasses, to claim that the postwar years or the men and women who lived through them were perfect. But it is only fair to note that the overwhelming lesson that I draw from World War II, and from knowing Ray Lambert, is not only that it was a remarkable time and he is a remarkable hero, but that all of us possess the capacity to do great things if we allow ourselves to be like him—humble, diligent, hardworking, and courageous when it is not easy to be.

—Jim DeFelice

Appendix A:

The Combat Medics of World War II

Combat medics in World War II carried a cloth medical bag or kit containing their essential supplies. There were several different versions during the war, differing slightly in size and things like the buckle, but all were army green and were essentially pouches with a fold-over top flap. They were held up by straps at the side, with a harness or suspenders, and slung over the body in different ways depending on personal preference, if photos from the period are any indication. The bags were usually near the medic's belt, where they were easy to open. The company aid men might have one or two of the bags; most of the others would usually carry only one, though at times like D-Day everyone had two pouches, stuffed with as much gear and supplies as they could get.

The items inside also varied slightly, depending on the year and place where they were deployed. A typical bag would include five yards of one-inch-wide surgical adhesive; scissors; forceps; safety pins; a small kit to treat burn injuries; boric acid ointment; eye dressing; iodine swabs; one-inch-by-three-inch bandages; morphine; sulfadiazine (or sulfa, as Ray calls it), and a tourniquet.

There were also medical tags to be used to record treatment data as well as personal information. Recording that a man had received a morphine shot was especially important to avoid overdoses.

Small personal items—cigarettes especially—were often carried as well.

Much more equipment was available in the large medical kits used at the battalion aid stations. Besides such staples as bandages, syringes, and the like, a battalion aid station would include supplies of common drugs like aspirin. Regimental aid stations, farther behind the lines, would generally have much more gear and equipment.

Combat medics like Ray and the men who worked with him were in a unique position. Not only were they shoulder to shoulder in combat with infantrymen on the front line, but they were the leading edge of a large operation that did most of its work by necessity far from the battlefield.

The army's medical department evolved after World War I, with changes due not only to medical advances but also to changes in the army's structure. During World War II, medical care was organized in a way familiar to the army, but sometimes difficult to understand for civilians not familiar with the 1940s command structure.

As members of the U.S. Army Medical Department, medics answered to a chain of command separate from the soldiers they treated. At the same time, their work meant they were "attached" to the parent unit of the men they took care of on the battlefield. They lived and trained with the infantry regiment, moved with it, fought with it. In a sense, they had two masters.

Even more confusing, at least for us, medics worked directly with medical units that were part of the regiment and division's regular chain of command, and which included doctors and other members of the medical department who were "assigned," rather than attached, to the division.

"Assigned" and "attached" have important meanings in the service, but while the difference was important to the men and the army, for us it's mostly irrelevant. Whatever hats or patches a man might wear, it's the job he does that's important.

The system followed the same general pattern throughout the war. A wounded soldier received first aid from a medic when he was wounded. The medic would decide if the man needed further treatment or not. If he did and he could not walk on his own power, stretcher bearers would take the wounded soldier to a battalion aid station. This station would typically be within a few hundred yards of the front lines—close walking distance, though in as safe an area as could be found.

Here, a doctor and other medics would give him more treatment. They might then send him back to his unit, or to a collection station.

This station would be a few miles farther back, and could receive men from different aid stations. More medical care could be administered here. Someone who was severely wounded would ordinarily go from the clearing station to a field hospital nearby, where complicated surgery could be performed. After that, or if his injuries had already been sufficiently stabilized, he would then go to a general hospital, either for more treatment or rehabilitation, called convalescence during the war. (Convalescent hospitals specializing in rehabilitation were also developed during the war.)

There were also evacuation hospitals, which received patients in less need of immediate care—often including shock victims and men with severe battle fatigue. Patients here would then go to general hospitals or replacement units, awaiting reassignment.

Care could be rendered at any stage, and it was always directed by members of the medical department who followed procedures developed by that department.

The collecting and clearing stations, the intermediate links in the system, were organized as part of the division itself. A battalion responsible for these stations—generally there would be two collecting and one clearing per division—would have about 460 men, including 34 officers and a chaplain. While called stations, these units and their facilities could and did render care as well as transport. The medical personnel there were part of the medical department, but they were assigned to the division as part of its organic units.

It's probably a lot more confusing on paper than it was in practice. The system brought treatment closer to the front lines and streamlined the process of getting the wounded to the hospital. The doctors especially seem to have moved around as needed. Overall, there would be roughly a thousand medical people in a division, with half in the assigned unit—technically called the medical battalion—and the other half in the medical detachment. Throughout the war, Ray was part of the latter.

What did the medical department treat?

Battle wounds get the most attention in histories of the department and the war, and for good reason: the most severe were acute and life threatening. But a good percentage of care involved what we might think of as mundane ailments, everything from a common cold to dysentery.

As Ray explains, trench foot and related diseases were an occupational hazard for troops who made seaborne landings in that era. Malaria and typhus were also a big hazard in combat, especially in the Mediterranean region, where sanitation was not advanced and the warm climate invited all sorts of problems.

Venereal disease was a major concern throughout the army, just as it has been probably since armies were first raised. Part of the treatment was education; the medical department produced numerous pamphlets and movies for the troops on a wide range of health topics. The general reception of these was about what you would expect, though probably no worse than the topics get today in the average high school health class.

There's no doubt that the medical department saved a lot of lives, far more than would have been expected had it not evolved after World War I and even the start of World War II. There were several primary reasons; medical advances were the most important.

Plasma, though sometimes impractical to administer and in short supply near the front lines, was a major lifesaver. It could be more easily transported and stored than whole blood, therefore allowing more patients to be stabilized or treated near the battlefield. X-ray machines allowed for better and quicker diagnoses. Rapid evacuation to what today we would consider advanced emergency trauma centers and to large surgical facilities improved survival rates as well.

The medical department's results were also due to increases in the amount of education medics and others received. When Ray first joined the army, medics were being taught little more than rudimentary first aid. His six-month stint in the Denver hospital was a relatively unique and new idea at the time. By war's end, schooling and training was specialized for medics, just as it was for "regular" infantrymen.

Soldiers who were wounded in battle weren't always in dire straits. Most probably didn't bother to report minor wounds. It was common for medics to shrug off shrapnel wounds that didn't keep them from doing their jobs, and the same was probably true for most other units.

If you were treated at the battalion aid station and didn't require hospitalization, there was a good chance that you would simply walk back to your unit. If you were taken for further treatment, though, you might have to be very lucky to make it back to that unit, even if your injuries weren't severe.

During much of the war, the army used a replacement system that would take soldiers who recovered from minor wounds and use them as replacements wherever they were needed. While from the command's point of view this made sense, from the soldier's point of view, it was one of the worst things that could happen. He had left his friends behind and now had to fit in with an entirely different group.

Many in the army railed against the system, including General Allen, who put his objections in writing.

Appendix B:

Battle Fatigue, Psychoneurosis, and PTSD in World War II

We tend to think of post-traumatic stress disorder or PTSD as a "modern" ailment, one afflicting soldiers and veterans in our current era, starting with Vietnam. But intense combat has affected warriors probably since the species evolved.

There are plenty of historical and literary accounts describing the effects of prolonged exposure to war. From the Sumerian epic Gilgamesh, written over four thousand years ago, to poems like Wilfred Owen's "Mental Cases," written in the middle of World War I, the effects of combat have been recorded, discussed, and debated for millennia. What is now diagnosed as PTSD is the most extreme example, but many soldiers whose symptoms do not rise to that level of diagnosis—which still carries a stigma—have had trouble readjusting to "normal" life after returning home.

In World War II, extreme cases of what were commonly called "battle fatigue" were classified as psychoneuroses by the army. Patients were extremely anxious and became "emotionally incapacitated" under stress. The men George Patton was alleged to have slapped would likely have fallen under such diagnoses.

The army's standard treatment of all battle fatigue in World War II was a return to techniques used in World War I. Psychiatrists at forward evacuation hospitals met and evaluated the men. The goal was to return them to full combat duty as quickly as possible; this was thought to help them recover. Various statistics show that, depending on the time and place, and the availability of beds, somewhere between 25 and 70 percent were returned.

During and after the war, the army studied psychoneurosis, trying to discover the cause. One of the most interesting findings was that many cases involved experienced NCOs—so many, in fact, that the researchers came to call it the "old sergeant's syndrome." An extraordinary number of experienced NCOs had broken down under the stress of continuous combat. While in many cases the symptoms were temporary—trembling during a shelling, being unable to make a decision or even move—the men had to be moved from their posts because they could no longer be counted on. Other symptoms included depression and survivor's guilt.

The only common factor seemed to be length of service in direct combat; ability or education before the onset of symptoms seemed to be irrelevant.

The studies after the war raised considerable doubts about whether sending the men back into combat really worked. But more advanced study and different treatment procedures lay well in the future. The words "battle fatigue" continued to carry a stigma; lesser cases were hardly ever recognized as a problem.

One factor that most of the early psychological studies may have missed was the effect of head injuries and concussions; to be classified as having such an injury, a man had to have extreme symptoms, including perforated eardrums and a "history of disturbed consciousness" and amnesia. Today, we know traumatic brain injuries often don't show such evidence.

Meanwhile, many soldiers had difficulty adjusting to civilian life after leaving the army and exhibited behaviors and problems that were surely due, at least in part, to the stress of combat. Ray's nightmares were typical of veterans who had been in heavy combat for extended periods. It was also common for soldiers to feel that they no longer fit into civilian life, and that America had changed in ways they didn't like. Veterans often resented civilians for not making what they thought were sufficient sacrifices, and for simply having no idea of what soldiers had been through during the war.

Eighty-one percent of the GIs surveyed by the Army Research Branch in 1945 reported that they had become more "nervous and restless" because of their service in the war. Discharged soldiers coped as well as they could, generally on their own. It was a time when, for the most part, talking about such problems would brand a man as weak. Nor did there seem to be much interest in hearing about the horrors of war.

Books and movies of the time generally glossed over the harsher realities. There were some notable exceptions, though. *The Best Years of Our Lives,* directed by William Wyler and released in 1946, was a box-office hit and won Wyler an Academy Award. Winning a total of seven Academy Awards, the film depicted the difficulties of three veterans returning from the war. Wyler had served as a major in the army air corps during the war. One of the supporting actors, Harold Russell, was an army veteran who had lost both arms in the war; he remains the only actor to have won two Academy Awards (one of them honorary) for the same performance.

Most veterans coped pretty well in the long run. While it's

become a cliché, the "Greatest Generation" remains a very accurate label when summarizing their accomplishments. Many of its members, including Ray, shrug when asked how they coped with the aftereffects of the war, or accomplished what they did.

"We didn't know any better," says Ray. "And we were all in the same boat."

His nightmares still come occasionally—the German soldier with the bayonet, artillery shells falling on the first aid station. Even seventy-five years after D-Day, the psychological effects of fighting World War II linger.

Ray is proof that a veteran can make a successful transition to civilian life. He thrived in business, raised a loving family, and was an important and giving member of his community. That is true of the overwhelming majority of veterans, from World War II to the present conflicts. But their success does not relieve the debt owed to combat veterans, nor relieve the country from the responsibility of helping that transition.

Appendix C:

Further Reading

One of our goals with this book is to keep interest in the war alive even as its participants pass on. No matter how thrilling, one man's account can only hint at the great expanse and drama of the historical event. We hope that Ray's story whets your appetite for more information. Here are a few places you can go to gain a broader perspective:

The First Infantry Division has an excellent museum a little over thirty minutes from Chicago at Cantigny Park in Wheaton, Illinois. The First Division Museum covers all of the Big Red One's history and has a lot of information and artifacts on World War II. More information can be found on its website: https://www.fdmuseum.org/

The Army Medical Department Museum is located at Fort Sam Houston in San Antonio, Texas. Check out the website, which includes a video tour: https://ameddmuseum.amedd.army.mil/.

The National World War II Museum in New Orleans, Louisiana, has artifacts, information, and programs on all aspects of the war; it's especially deep in the area of D-Day. The website: https://www.nationalww2museum.org/.

The National D-Day Memorial in Bedford, Virginia, is an inspiring memorial to the soldiers who fought on the beaches and often features programs with surviving World War II veterans. Bedford and the surrounding area was the home of many of the members of the 116th Infantry Regiment, a National Guard unit that was part of the 29th Infantry and landed at the same time Ray did. https://www.dday.org/.

The 16th Infantry Regiment Association maintains a website: https://16thinfassn.org/.

Immediately following the war, the U.S. Army produced an official history of operations; currently available online, it remains an accessible and informative introduction for general readers. A general readers' guide to these volumes and others associated with World War II can be found at this website: https://history.army.mil/books/wwii/11-9/11-9c.htm.

There are countless books, movies, and videos on the war and D-Day in particular.

Rick Atkinson's Liberation Trilogy follows the American army through the European theater. The first book, *An Army at Dawn,* details the North Africa campaign. *The Day of Battle* details Sicily and the fight in Italy. *The Guns at Last Light* takes the reader through Normandy, France, and into the end of the war in Germany. *The Mighty Endeavor* by Charles B. MacDonald is an older book that covers the same ground in one volume.

Among the multitude of books meant for general readers on D-Day, Stephen Ambrose's *D-Day: June 6, 1944: The Climactic Battle of World War II*; *Overlord* by Max Hastings; and the longer *D-Day: The Battle for Normandy* by Antony Beevor stand out.

Though older than the others, *The Longest Day* by Cornelius Ryan is another great one-volume read. A 1962 film version is considered one of the greatest war films of all time. The more recent *Saving Private Ryan,* while fictional, includes scenes of the D-Day landings that many veterans consider highly realistic.

The Dead and Those About to Die by John C. McManus details the 1st Infantry Division's actions on D-Day and is a valuable starting place for learning about the famed unit's most dramatic World War II battle.

Notes and Sources

GENERAL SOURCES

In addition to interviews and discussions between Ray Lambert and Jim DeFelice over a period of several months, many additional sources provided important background, context, and additional facts and perspective. Among the most significant:

- Some years ago, Ray recorded his memories with the help of a neighbor. Some of those transcriptions provided information for questions and background for our interviews.
- Those interviews and others were the basis for an account written by Ray with assistance from Colonel (Ret.) Christopher D. Kolenda, Ph.D., which also provided prompts for questions for our interviews.
- Over the years, Ray has been the subject of a large number of news stories, both in print and other media; these also were very useful background and helped inform our conversations.
- Members of Ray's family provided additional information.

ADDITIONAL SOURCES

Records of the 1st Division, especially after-action reports, and histories of the First Division, the 16th Infantry, and the Medical Department.

Formal and informal interviews with other veterans.

Army Signal Corps, Navy, and Coast Guard original photos and film footage of the war.

Major Charles E. Tegtmeyer, Personal Wartime Memoir Historical Unit, Army Medical Department in 1960, located in the National Archives and Records Administration, Record Group 319. Excerpts are available at: https://history.amedd.army.mil/booksdocs/wwii/Normandy/Tegtmeyer/TegtmeyerNormandy.html.

A number of books provided important background, helping to understand the context of the battles Ray participated in, including all of the works mentioned above in "Further Reading," Appendix C, plus:

The official U.S. Army history of the war, *United States Army in World War II*, CMH Publication 11-9, Center of Military History, United States Army, Washington, D.C., 1992. Available online at: https://history.army.mil/html/bookshelves/resmat/ww2eamet.html.

Joseph Balkoski, *Omaha Beach: D-Day, June 6, 1944*, Stackpole Books, 2004.

James Jay Garafano, *After D-Day: Operation Cobra and the Normandy Breakout*, Lynne Rienner Pub., 2000.

Norman Gelb, *Desperate Venture: The Story of Operation Torch*, William Morrow, 1992.

Kent Roberts Greenfield et al., *The Organization of Ground Combat Troops*, Historical Division of the Army, 1947.

Orr Kelly, *Meeting the Fox: The Allied Invasion of Africa, from Operation Torch to Kasserine Pass to Victory in Tunisia*, John Wiley & Sons, 2002.

David M. Kennedy, editor, *The Library of Congress World War II Companion*, Simon & Schuster, 2007.

Cole C. Kingseed, *From Omaha Beach to Dawson's Ridge: The Combat Journal of Captain Joe Dawson*, Naval Institute Press, 2005.

Leo Marriott and Simon Forty, *The Normandy Battlefields*, Casemate, 2014.

Charles Messenger, *The D-Day Atlas: Anatomy of the Normandy Campaign*, Thames & Hudson, 2014.

Charles Shay, *Project Omaha Beach: The Life and Military Service of a Penobscot Indian Elder*, Polar Bear & Company, 2012.

Colonel Roy M. Stanley II, *The Normandy Invasion, June 1944: Looking Down on War*, Pen & Sword Military, 2013.

Flint Whitlock, *The Fighting First: The Untold Story of the Big Red One on D-Day* (Kindle Edition), Westview Press, 2005.

Additionally, several visits to France, the battlefields and museums there, most recently in October 2018 when Ray's Rock was dedicated, helped orient the accounts of D-Day and its aftermath. Interviews with museum staff and guides were also extremely helpful.

NOTES TO CHAPTERS

ONE

The price of milk is drawn from "Wages During the Depression" by Leo Wolman, published in 1933 by NBER. A pdf version is available on the web at https://www.nber.org/chapters/c2256.

TWO

Harmony Church remains part of Fort Benning; it is currently home of the Armor School. The World War II–era barracks were used by soldiers into the twenty-first century.

Declassified documents detailing the 1st Division's activities during 1941 and 1942 before the deployment to England include the following highlights:

- The division "consolidated" or brought together its regiments and other units at Fort Devens during the winter of 1940, adding an antitank battalion (1st Antitank Battalion, which became the 601st Tank Destroyer Battalion).
- Large-scale training exercises included landings at Buzzards Bay in Massachusetts, New River, North Carolina, and Puerto Rico.
- A series of landings were conducted in North Carolina at New River from June through August 1941.
- In January 1942, the division practiced landing at Virginia Beach.
- In February 1942, the division moved to Camp Blanding in Florida for extensive training, then moved up to Fort Benning. In June, it was relocated to Indiantown Gap, Pennsylvania, preparing to go to England.

In addition to the 16th, 18th, and 26th Infantry Regiments, four field artillery battalions, the 5th, 7th, 32nd, and 33rd, along with the 1st Engineers, were part of the 1st Division during the war.

Force numbers for the prewar and war years come from *The Organization of Ground Combat Troops* by Kent Roberts Greenfield, Robert R. Palmer, and Bell I. Wiley, part of the U.S. Army in World War II collection (CMH Pub 2-1, 1947, 2004).

Among other things, Maxwell hosted the Army Air Corps Tactical School at the time Ray joined the service, and played a leading role training pilots in developing the tactics they would take to war. Information on the field's prominence rests in part on "Historical Picture of Maxwell AFB" by Major Larry Edward Kangas, a student at the Air Command Staff College, prepared in April 1986. The unclassified report is available online at https://apps.dtic.mil/dtic/tr/fulltext/u2/a168255.pdf.

Some of the 1st Division's history comes from the First Division Museum and its website: https://www.fdmuseum.org/.

Besides the three regiments cited and a headquarters company, the 1st Division in World War II included the 1st Reconnaissance Troop (Mechanized),

1st Engineer Combat Battalion, 1st Medical Battalion, 1st Division Artillery, 7th Field Artillery Battalion (105mm Howitzer), 32nd Field Artillery Battalion (105mm Howitzer), 33rd Field Artillery Battalion (105mm Howitzer), 5th Field Artillery Battalion (155mm Howitzer), 701st Ordnance Light Maintenance Company, 1st Quartermaster Company, 1st Signal Company, a Military Police Platoon—and a band. A wide variety of units, mostly artillery, armor, and anti-air, were "attached" during the course of the war, most for very short periods of time.

The details on Bill Lambert's initial assignment with the cavalry and why he was at Fort Benning are not entirely clear. At the time, the United States was transitioning to armor and mechanized units, and it's possible that had he not become a medic and joined the 16th Regiment, he would have ended up in an armor outfit.

Battalion aid stations, where Ray spent much of his time supervising the unit, were generally located within a mile of the front; they were often much closer. The next stop for a wounded man was a collecting station, generally located with or near the regimental headquarters. Patients there would go directly to a hospital or a clearing station located a few miles farther back. At that point, they would be transported to a field hospital or directly to a permanent general hospital for further treatment.

The overall flow was meant to consolidate transportation while quickly caring for the most seriously wounded. The system moved advanced treatment far closer to the front lines than in World War I; that and advances in medicine helped save many men who would have died in the earlier war.

At any point in this chain, the wounded might receive further treatment or be cleared to return to combat. There were different contingencies depending on time, place, and circumstance, but the general outline was followed throughout the war in all theaters.

Ernie Pyle's description is from a column dated December 30, 1940. Entitled "A Dreadful Masterpiece," it talked about the horrible and extremely ironic beauty of the war seen from a distance—an apt description of what people were experiencing, though that was not Pyle's intent. The column is collected in numerous places on the internet, including here: http://mediaschool.indiana.edu/erniepyle/1940/12/30/a-dreadful-masterpiece/ (In print: *Ernie's War: The Best of Ernie Pyle's World War II Dispatches,* edited by David Nichols, pp. 42–44.)

Fort Jay and Governors Island in New York Harbor played an important role in the U.S. Army from the time of the revolution. At the time of Ray's assignment there, army and air corps headquarters were located there; it housed First Army command during the beginning stages of the U.S. involvement in World War I. The 16th Infantry had first been posted to Jay in 1922. It is now part of the Governors Island National Monument, with a park and historical buildings open to the public.

Ray was stationed there from roughly the beginning of June 1940 to the end of February 1941, with at least one substantial period of maneuvers in upstate New York in August 1940.

The 1941 version of "Green Eyes" is available on YouTube at https://youtu.be/V8hMyVQaCUo.

Fort Devens has been part of the U.S. Army since World War I; it can trace its military roots even farther back. Over the years, it has alternatively been known as Camp Devens, with Devens occasionally styled as Deven.

The installation was named for Massachusetts general Charles Devens, who served with distinction during the Civil War, rising from the rank of major, and later became a judge. The base was closed in the 1990s, but part of the property is currently used as the Devens Reserve Forces Training Area.

THREE

The full recording of the news flash on Pearl Harbor is available at the National Archives, and online at https://catalog.archives.gov/OpaAPI/media/2192572/content/arcmedia/mopix/audio/ww2/200-54.mp3.

The reporter's tone is remarkably calm and businesslike; not all were.

As is often the case with World War II battles, the exact casualty rate at the Battle of Gazala is in question. Historians have used a wide range of numbers when estimating those killed, wounded, and captured; 35,000 is at the high end of those estimates. The defeat led to a change in command on the British side, where eventually its superior numbers would overmatch Rommel's superior generalship.

Tobruk had been successfully defended by the British the year before in an epic battle and siege.

The army technician ranks went from 5 to 3; the lower the number, the higher the rank. While T-3 is often compared to a staff sergeant's rank, it did not carry the full authority of a staff sergeant.

The information on convoy AT-16 comes from the Arnold Hague Convoy Database.

The *Bedford* served throughout the war; refitted, she sailed until 1960.

"Drunken Duchess" comes from a description in *John Buchan: A Memoir,* by William Buchan, published by Buchan & Enright, 1982.

The incident with the *Bedford* and its U-boat has been of interest to many World War II and naval buffs for some time. Definitive proof of a U-boat being sunk, either by a convoy escort or later by the troopship herself, is lacking.

Given the limited technology of the time, it was easy to mistakenly believe a submarine was sunk in action when it had merely been run off; convoy reports and eyewitness accounts are filled with such examples.

Some of the description of the *Bedford* relies on period postcards and an account by J. E. Robson, an RAF member whose story was published by the BBC in June 2005.

FOUR

Aside from the official and general sources previously cited, background and information on Torch comes from First Division after-action reports; Gray, Tavares, et al., *CST Battlebook 3-A Operation Torch,* Combat Studies Institute, Fort Leavenworth, Kansas. (Available as pdf here: https://apps.dtic.mil/dtic/tr/fulltext/u2/a151625.pdf); Norman Gelb, *Desperate Venture: The Story of Operation Torch* (New York: William Morrow, 1992); Vincent O'Hara, *Torch: North Africa and the Allied Path to Victory* (Annapolis: U.S. Naval Institute Press, 2015); and Charles Moran, ed., *The Landings in North Africa: November 1942* (Washington, DC: Publications Branch, Office of Naval Intelligence, United States Navy, 1944). Additional sources included *Algeria-French Morocco* by Charles R. Anderson, published by U.S. Army Center of Military History and available at https://history.army.mil/html/books/072/72-11/CMH_Pub_72-11.pdf; *Operation Torch: The American Amphibious Assault on French Morocco, Naval History and Heritage Command,* available in pdf format at https://www.history.navy.mil/content/dam/nhhc/research/publications/Operation-Torch-booklet-508.pdf.

Ray remembers boarding the *Bedford* on October 15, a date backed up by the ship's records. The starting date of the convoys south from England usually given in general and official histories is October 22.

Dr. Morchan passed away in 1972 at the age of fifty-nine. He was a consulting radiologist at an Indiana hospital at the time and was highly regarded by his colleagues for his skills and his friendly demeanor.

During the war, regiments often operated as the core and command unit for groups of smaller units temporarily attached to them. As such, they were officially referred to as "Regimental Combat Teams" (RCT) or sometimes just "Combat Teams," with the command regiment's number—16th RCT, for example. For simplicity's sake, we generally refer to the regiment rather than RCTs in the book.

The typical RCT during the Africa landing included one regiment, one medical detachment, one company of field engineers, a battalion of field artillery, a signal (communications) detachment, an engineer battalion specializing in landing and shore operations, and an anti-air battalion. Other units, armor especially, were often attached depending on specific missions.

This general arrangement continued through the war.

The unit history indicates the company arrived on the beaches east of Oran earlier than Ray remembers, around 10 A.M.

Besides the earlier sources, some of the description of the battles around Oran benefit from the account and photos in "The Landing at Oran" by Ben Hilton, posted on the 16th Infantry historical site, available at https://www.16thinfantry.com/unit-history/the-landing-at-oran/.

For vital statistics on the 88 as well as its fame, see Ian V. Hogg, *German Artillery of World War Two,* 1997. Hogg makes clear that the gun's fame among Allied soldiers was as much due to its ubiquitousness and propaganda as its actual abilities, which were considerable.

Information on landing conditions and lessons learned in Africa comes from a declassified after-action report for Allied Force Headquarters. Interestingly, those reports lambaste the Higgins boats, labeling them inadequate in many respects, from size to armor. While the army introduced a large number of other landing craft with different missions and significant improvements, the Higgins boats were used on the subsequent Sicily and Normandy invasions, with great success.

Medical Detachment Jeeps were used for a variety of purposes, serving not only as scout and basic transport vehicles, but also as makeshift ambulances. Depending on what was available, units would sling pipes and other items across to accommodate up to five stretchers; the wounded could be carried across the hood as well as the backseat area. The only limits to the adaptations were imagination and available materials.

Patton secretly left Africa around the end of March, which meant Bradley mapped out the last offensive and was actually in charge for the final victory.

The sequence of events following the American defeat at Kasserine Pass has been reconstructed based on unit medical and Purple Heart records.

FIVE

McNair received a Purple Heart for the injury treated at Ray's aid station; the date was April 22, 1943.

Lieutenant Colonel Denholm later won the Distinguished Service Cross for bravery during a battle in Sicily in July when he and two other officers and two enlisted men held off a number of tanks so the rest of his men could safely retreat. He was wounded in that battle.

Hill 609 was eventually taken in a tank and infantry attack ordered by Bradley on April 30. The plan was both highly unorthodox and extremely risky, but it worked. When the scorched terrain—the 34th had fired phosphorous shells during the battle to smoke the Germans out—was cleared, the hill belonged to the Americans.

Historian Rick Atkinson described the summit as "hell's half-acre."

Different versions of what happened aboard the Italian ship circulated after the incident, and some credited the prisoners with sailing the ship back to shore. That appears not to be true.

SIX

Besides the general sources cited above, the overall description of the battles relies heavily on the First Division after-action reports. The following sources were useful for background and understanding of the overall battle:

Rick Atkinson's *The Day of Battle*; Albert N. Garland and Howard McGaw Smyth, *Sicily and the Surrender of Italy*; Carlo D'Este, *Bitter Victory, The Battle for Sicily, 1943*; S. W. C. Pack, *Operation HUSKY: The Allied Invasion of Sicily*; Samuel Eliot Morison, *Sicily-Salerno-Anzio*.

The *History of the Sicilian Invasion* by Andrew J. Birtle, U.S. Army Center of Military History (available at https://history.army.mil/brochures/72-16/72-16.htm), provides a succinct summary of the battle's progress.

One often-overlooked aspect of the landings at Sicily—and Normandy, for that matter—is that the U.S. Coast Guard also supplied boats, material, and personnel for the landings.

The 1st Fallschirm-Panzer Division Hermann Göring, the official name of the Hermann Goering Division, was probably overrated by Allied intelligence at least partly because of its name and relationship with the important German figure. Nonetheless, its ranks included well-trained and battle-tested paratroopers and experienced tank crews imported from other units. The division also was among the most notorious for war crimes.

The Terry Allen quote is taken from Rick Atkinson's book.

The casualty figures come from *U.S. Army in World War II, Sicily and the Surrender of Italy,* by Garland and Smyth.

At the time of the 1st Division attack on Troina, the 9th Infantry Division had come onto the island and was due to replace the Big Red One.

We had difficulty determining when the tank incident took place. Ray has placed it here after consulting with Steven Clay, historian and president of the 16th Infantry Regiment Association.

For a different take on Terry Allen's firing, see "Investigation into the Reliefs of Generals Orlando Ward and Terry Allen," by Major Richard H. Johnson, published as a monograph by the School of Advanced Military Studies, United States Army Command and General Staff College, Fort Leavenworth, Kansas, 2009. (Available as a non-classified pdf: https://apps.dtic.mil/dtic/tr/fulltext/u2/a505159.pdf.)

Johnson argues that Eisenhower planned to move, not fire, Allen all along.

There are admittedly much harsher versions of the slapping incident from different eyewitnesses, who said Patton inappropriately lost his temper. See, for example, Atkinson's version on pages 147–49 in *The Day of Battle.* Patton's own journal entry backs up the harsher views.

As was common when the medals were awarded in the field, Ray did not receive a citation at that time for what would have been his second Silver Star (technically, an oak cluster). The paperwork was apparently lost after the war. Because he does not have the documentation, Ray does not list it among his official awards.

SEVEN

The men who died in Exercise Tiger at Slapton Sands were part of what would have been a follow-on force after the 4th Infantry Division landed on Utah Beach. The disaster eliminated the LSTs that would have been part of the reserve for D-Day. The impact of the accident is difficult to assess, but surely significant.

Over the years there have been charges of a "cover-up" relating to Slapton Sands. Those appear to be misguided. See, for example, "Slapton Sands: The Cover-up That Never Was" by Charles B. MacDonald, *Army* 38, No. 6 (June 1988): 64–67.

Dawson's notes about the court-martial are included in *From Omaha Beach to Dawson's Ridge* by Cole C. Kingseed, Naval Institute Press, 2003.

Information on the order of battle, commanders, etc., is from *Omaha Beachhead (6 June–13 June 1944), American Forces in Action Series,* Historical Division War Department (Facsimile Reprint, 1984, 1989, 1994, CMH Pub 100-11, Center of Military History), Washington, D.C.: United States Army. (Available on the web at https://history.army.mil/books/wwii/100-11/100-11.HTM.)

The medical department's plans for evacuation and treatment of the wounded were summarized in a newsreel prepared by the army during the war and can be seen at: https://www.youtube.com/watch?v=GXj-DCZMb7o.

EIGHT

The machine gun nest on the left that Ray remembers would have been on the hill, roughly in line with the Engineers' monument and what is now "Ray's Rock"—the monument to him and the other medics who served on the beach. The right machine gun would have been on the hill below what is now the American Cemetery.

There are many versions of E Company's heroics as they pushed inland. For a concise, well-written account, see John C. McManus's "A Knife in the Vitals: Omaha Beach," published in *World War II Magazine*, February 15, 2017, and available online at: https://www.historynet.com/knife-vitals-omaha-beach.htm.

Ray Lepore's obituary appeared in the July 10, 1944, edition of the *Boston Globe.*

TEN

Some Sherman flail tanks had their mine-killing mechanism at the rear of the tank rather than the front.

The medal citations are collected at https://ameddregiment.amedd.army.mil/dsc/wwii/wwii_ad.html.

Ray didn't know it, but there were at least two other Lamberts in the division: Tech 5 Ross E. Lambert, who was in the 1st Medical Battalion, and Clyde C. Lambert, who was in Company I. Clyde received the Bronze Star for action at Normandy. Ross died during the war; it's not clear from division records when or where he died.

Even now, casualty figures for D-Day vary slightly from source to source. First Army, the parent of 1st Division, counted a total of 1,465 men killed, with another 1,928 and 6,603 wounded in the first twenty-four hours of the operation. Looking at the overall Battle of Normandy, usually dated from June 6 to the completion of the German retreat over the Seine on August 30, there were over 209,000 Allied casualties, with 37,000 dead, not counting nearly 17,000 airmen who died in the same period. German casualties were far greater, numbering well over 200,000, a significant portion dying or taken prisoner during the Falaise Pocket battles in August.

The U.S. Army had nearly 600,000 casualties in battle during the war; over 26,000 were killed. More than half of the wounded returned to duty, though not always to their own units or the front lines.

ELEVEN

Ray has the Silver Star medals he was awarded, but the whereabouts of the documents for numbers two and three remain a mystery. Neither is recorded in the documents he received in the 1990s when requesting a review of his medals and discharge paperwork.

Like that of so many others, his original paperwork was lost in a fire that consumed government records at the National Personnel Records Center in St. Louis in 1973; the fire destroyed about 80 percent of the records of U.S. Army servicemen discharged between 1912 and 1980. Though some of these records were reconstructed from other sources, large gaps similar to those in Ray's record remain.

Officially, his service records indicate one Silver Star and two Bronze Stars with oak clusters (effectively, four Bronze Stars total), as well as a number of other awards.

For more information on the fire and records recovery efforts, see the post at: https://www.archives.gov/personnel-records-center/fire-1973.

Ray wears a single Silver Star on his uniform, and generally refers to that award in the singular when asked. It is only after prompting that he tells the story of the lost medals.

TWELVE

Euel W. "Bill" Lambert's obituary states he was born October 19, 1917, and died July 16, 2010.

Acknowledgments

Ray has a large and generous group of friends and family, and their love and support truly made this book possible. Thanks first and foremost to Barbara, Ray's wife, who helped with a variety of tasks directly related to the book and was a gracious, patient, and welcoming host as the interviews progressed. As Ray notes in the book, she's the reason he's still going strong at ninety-eight.

Besides being great friends, Ray's neighbors, Gijsbertus Van-Domselaar, otherwise known as Bert, and his wife, Elly, were a tremendous help to him when he was first organizing his memories.

Ray would also like to specially thank Major Christophe Coquel of the French Gendarmerie. Christophe has gone out of his way to make things easy for Ray's visits to Normandy, and had been a tireless supporter of the 16th Infantry and the Big Red One for many years.

Critical research on this book was provided by Ray Fashona, an excellent writer as well as researcher, who did much of the footwork in the preliminary stages, tracking down survivors of the Battle for Normandy and their families.

Debra Scacciaferro was again of immense assistance, researching different aspects of the battles and medical history. She also spent considerable time sorting through photographs and official U.S. Army documents to augment my research. In addition, Debra provided valuable feedback as the manuscript was being prepared.

In France, Jim's friend Dianne Condon-Boutier (now with Fat Thelma Tours, www.france-vacations-made-easy.com) was a dynamic, tireless, and truly phenomenal guide, translator, and sounding board in Normandy, literally opening doors to hidden bunkers and normally closed museums. Jim's friend Patrick Ober was a great companion and resource, as well as a cheerful translator and sounding board.

Stéphane Leguennec of La Colline was a gracious and accommodating host in Bayeux. Label West Tours helped facilitate Jim's arrangements with great efficiency.

Many museum and library directors and staffs played an important role in providing background and information. John Long at the National D-Day Museum was especially helpful and extremely gracious as a host during the early stages of research on the book. Jim was assisted once again by the Ramapo Catskill Library System and the staff at Albert Wisner Library, who obtained many difficult-to-find books that added to this project.

Our editor at William Morrow, Peter Hubbard, was instrumental in the birth of this book. Not only did he suggest the topic, he tirelessly pushed and supported it, and accompanied Jim on an early trip to work with Ray.

Peter's assistant, Nick "Cadillac" Amphlett, was a critical resource at the publishing house and while we were doing research.

Vice President of Publicity Sharyn Rosenblum and her staff were wonderful in their support.

Also at William Morrow, sincere thanks to President and Publisher Liate Stehlik for her early enthusiasm and continuing support; art director Rich Aquan for his creative work, Judy R. DeGrottole, Nyamekye Waliyaya, and Andrea Molitor.

Index